Every Last Drop

Praise for this book

'In terms of its scale, its long-term approach and the severity of the environmental constraints it had to face, the Gansu water programme is of global significance. The technologies and approaches used will be of interest to many outside China, even if the conditions they face are less severe. This book is a major source of tested techniques in the fields of water supply and agriculture for semi-arid lands, solar energy and erosion control. I happily commend it.'
Dr Terry Thomas, Head Development Technology Unit, Warwick University, UK, and author, researcher and consultant in tropical rainwater harvesting

'*Every Last Drop* shows that great strides can be made against the severe challenges of water shortage and soil conservation, without large expenditures, by using the best of traditional technologies combined with modern science and innovation. Millions of people are living better lives thanks to the sometimes simple but effective techniques of capturing rainwater for domestic use and agriculture. Required reading... and a strong case for more demonstration projects!'
David Bainbridge, Agroecologist, California

'This book is an important addition to the literature on rainwater harvesting. It demonstrates in simple and understandable terms how the use of rainwater both for domestic needs and agriculture has developed in Gansu Province and is practised today. There are very many comparable situations all over the world and we all can learn much from Gansu. Now, not only those who have seen with their own eyes but also readers of this book will turn into believers of the huge, yet untapped potential of rainwater harvesting.'
Hans Hartung, expert in rainwater harvesting

Every Last Drop
Rainwater harvesting and sustainable technologies in rural China

Zhu Qiang, Li Yuanhong, and John Gould

Practical Action Publishing Ltd
The Schumacher Centre
Bourton on Dunsmore, Rugby,
Warwickshire CV23 9QZ, UK
www.practicalactionpublishing.org

© Practical Action Publishing, 2012

ISBN 978 1 85339 737 0 Hardback
ISBN 978 1 85339 738 7 Paperback

All rights reserved. No part of this publication may be reprinted or reproduced or utilized in any form or by any electronic, mechanical, or other means, now known or hereafter invented, including photocopying and recording, or in any information storage or retrieval system, without the written permission of the publishers.

A catalogue record for this book is available from the British Library.

The authors have asserted their rights under the Copyright Designs and Patents Act 1988 to be identified as authors of this work.

Since 1974, Practical Action Publishing (formerly Intermediate Technology Publications and ITDG Publishing) has published and disseminated books and information in support of international development work throughout the world. Practical Action Publishing is a trading name of Practical Action Publishing Ltd (Company Reg. No. 1159018), the wholly owned publishing company of Practical Action. Practical Action Publishing trades only in support of its parent charity objectives and any profits are covenanted back to Practical Action (Charity Reg. No. 247257, Group VAT Registration No. 880 9924 76).

Cover photo: Terraces in Zhuanglang County, Gansu ('China's terrace county') have provided multiple benefits for agriculture, the environment, and the local economy through conserving soil and water.
Photo credit: Q. Zhu, 2009
Typeset by S.J.I. Services
Printed by Hobbs the Printers Ltd
Printed on FSC mixed source paper

Contents

Tables, figures, photos, and boxes	vii
Acronyms and abbreviations	xii
About the authors	xiii
Preface	xv
Acknowledgements	xvii
Introduction	1
PART I – Rainwater harvesting and domestic rural water supply	**3**
1. Development of rainwater harvesting in rural Gansu	5
Gansu climate and water resources	5
Rainwater storage tanks	11
Traditional *Shuijiao*	11
2. Upgraded water cellar designs	15
Thin-walled water cellar	15
Concrete domed water cellar	17
Cylindrical water cellar	19
Rainwater harvesting system design procedure and development	21
Replication of new rainwater harvesting techniques in Gansu	24
Reasons for the rapid spread of rainwater harvesting systems	25
3. Water quality and the development of solar cooker technology	29
Water quality issues	29
Technical aspects of the solar cooker	30
Proper use of the solar cooker	37
Popularization of the solar cooker in rural Gansu	38
Case study: Qiang Li and his Huaneng Solar Energy Co. Ltd	38
PART II – Rainwater harvesting and sustainable agriculture	**41**
4. Development and replication of low-cost greenhouse designs	43
Background	43
General introduction to the greenhouses in Gansu	46
Design features of the greenhouse	50
5. Development of rainwater-harvesting-based irrigation systems in Gansu	59
The challenge of dryland farming and some innovative solutions	59
Rainwater-harvesting-based irrigation systems	62
Principles and feasibility of rainwater-harvesting-based irrigation	67

6. Irrigation methods using rainwater 71
    Simple and affordable irrigation methods 71
    Micro-irrigation systems 78
    Micro-catchments using rainwater concentration 85
    Optimizing the irrigation schedule 88

PART III – Rainwater-harvesting and environmental management 93

7. Small watershed management 95
    Background 95
    Scientific approach to soil erosion control 96
    Terracing 100
    Growing trees and grass 108
    Bio-engineering measures for soil erosion control 113
    Impacts of small watershed management in Gansu 118

PART IV – Challenges, future prospects, and conclusions 125

8. Challenges and prospects for rainwater harvesting in Gansu Province 127
    Challenges of rainwater harvesting for domestic water supply 127
    Challenges of supplementary irrigation with rainwater harvesting 128
    Prospects for the future of rainwater harvesting 130
    Conclusions 131

Appendix 1 Water purification system for harvested rainwater in Gansu 133

Appendix 2 Solar cooker design details: Parabolic curve and how to draw it 137

Notes 139

References and further sources of information 141

Tables, figures, photos, and boxes

Tables

1.1	The precipitation and water resources in the world, China, and Gansu (2003)	5
2.1	Unit cost of different upgraded *Shuijiao* (CNY)	20
2.2	Yearly mean RCE of different types of catchment (%)	23
2.3	The volume coefficient K in Equation 2.2	23
2.4	Dimensions of circular water cellars	24
4.1	Irrigation water use for different irrigation methods and crops	45
4.2	Yields and values of different crops from the greenhouse	46
4.3	Yield versus quantity of irrigation for cucumber crop	46
4.4	Rainwater collected from the greenhouse roof and annual water demand	49
4.5	Span of greenhouse related to the outdoor lowest temperature	51
4.6	Height versus span of the greenhouse	51
5.1	Irrigation frequency and amounts for rainwater harvesting systems	67
5.2	Application frequency and quota for rainwater harvesting irrigation	67
5.3	Results from testing and demonstration projects	68
6.1	Effect of irrigation during seeding on yield of corn	73
6.2	Water productivity of irrigation during seeding of corn	73
6.3	Specification of drip line produced in China	80
6.4	Equipment for movable hand pump drip system for 2 mu of field crop	82
6.5	Equipment for movable electric pump drip system for a 4 mu field crop	83
6.6	Irrigation area and sprinkler intensity versus number of sprinklers	84
6.7	Water deficit in the growing period for the main crops	85
6.8	Results of r versus annual rainfall as recommended by GRIWAC	86
6.9	Yield increases by adopting micro-catchments in corn and potato tests	87
6.10	Effect of plastic sheeting after harvest on soil moisture conservation	87

viii EVERY LAST DROP

6.11	Spring wheat yield by irrigation amount in the growing season	89
6.12	Spring wheat yield under different irrigation schedules	89
6.13	Percentage yield and WSE increases for two applications compared to one application at either jointing or booting	90
6.14	Corn yield under different irrigation schedules	91
6.15	Millet yield by irrigation timing	92
6.16	Yield of millet by irrigation quota	92
7.1	Results for optimized terrace parameters	103
7.2	Runoff and soil erosion rates for different vegetation structures	112
7.3	Size of earth check dams	116
7.4	Classification of silt-retaining dams in China	117
7.5	Flood control standard for silt-retaining dams	118
7.6	Reduction of silt discharge versus coverage of control	119
7.7	Comparison of flood discharge between Qiangjiagou and Tianjiagou watersheds	120
7.8	Runoff reduction versus percentage of erosion control and silt reduction	121
A2.1	Recommended focal point distances for solar cookers	137

Figures

1.1	Map of China showing Gansu Province	6
1.2	Traditional *Shuijiao* water cellar using clay-mud paste for seepage control	12
2.1	The water cellar with top semi-dome structure	16
2.2	Construction procedure for building water cellar with top dome structure	18
2.3	The water cellar with cylinder-structure wall	19
2.4	Cost of *Shuijiao* (unit cost versus volume)	20
3.1	Reflection of parallel light beams from parabolic curve	31
3.2	Template of parabolic curve	35
4.1	Arch-type plastic greenhouse	47
4.2	Solar-heated walled greenhouse	47
4.3	Components of the walled greenhouse	48
5.1	Water demand of spring wheat versus natural rainfall in a normal year	60
6.1	Cross-section of multi-function sowing machine	72
6.2	Illustration of furrow irrigation under plastic film	75
6.3	Water injector	76

6.4	Seepage irrigation for fruit trees and corn using porous pots: (a) four vessels around a tree; (b) three corn plants per vessel; and (c) a percolation vessel	77
6.5	Illustration of mini-drip irrigation system	78
6.6	Screen-type filter	79
6.7	Layout of hand pump drip system	81
6.8	Layout of fixed drip system for 400 m² greenhouse	83
7.1	An arrangement of erosion control measures in Qianjiagou watershed (representing areas of loess plateau with gullies)	98
7.2	An arrangement of erosion control measures in Baozigou watershed (representing areas of loess hills with gullies)	98
7.3	Two types of terrace: (a) continuous terracing and (b) terraces with the original slope partly retained	101
7.4	Section of cut and fill-in terracing	104
7.5	Cascade group of check dams	115
A1.1	Flow chart of the purification system	134
A1.2	Section of the filter	134
A2.1	Drawing the parabolic curve	141

Photos

1.1	Soil erosion and cultivation on the steep land are both common features of the Loess Plateau in Gansu Province	7
1.2	Rainwater harvesting using tiles and concrete paving, for domestic water use	9
1.3	Rainwater harvesting for supplemental irrigation	10
1.4	Rainwater harvesting systems for ecological conservation	10
2.1	A simple platform enables the building of a brick semi-dome for the top of a water cellar	19
2.2	Water delivery facility using a hand pump and a bucket and rope	21
3.1	Solar cooker with type I frame	33
3.2	Solar cooker with type II frame	33
3.3	Mould with cast of the cooker body	36
3.4	Plastering the first layer of mortar	36
3.5	Sticking on the mirrors	37
3.6	The completed solar cooker bodies	37
3.7	Foldable solar cooker	39
4.1	Plastic shed – the first greenhouse in the agricultural history of Beishan mountains in Yuzhong in 1995	44

x   EVERY LAST DROP

4.2	Group of greenhouses in Dingxi County	50
4.3	Greenhouse without columns	53
4.4	Cover made of plastic tarpaulin	56
4.5	Device for furling covering material	56
4.6	Greenhouse wall built with polystyrene insulation board	56
4.7	Two-span greenhouse	57
5.1	Rainwater harvesting for irrigation using a highway for the catchment	62
5.2	The highway passing Bao's house becomes his reliable water source for raising crop yield and improving his life	63
5.3	The water collected from the highway used to irrigate Bao's greenhouses, his field, and the garden	64
5.4	Concrete-lined catchment	64
5.5	Greenhouse roof as catchment	64
5.6	Tank by the side of cornfield	65
5.7	Group of tanks along a canal	65
5.8	Surface irrigation tank	65
5.9	Water cave for storage	65
5.10	Hand pump for field irrigation	66
6.1	Integrated machine for sowing and laying plastic sheeting	71
6.2	Comparison between (a) with and (b) without irrigation during seeding	73
6.3	Watering crops through holes on the plastic film	74
6.4	Manual irrigation using a hose	77
6.5	Built-in emitter in the drip line	80
6.6	Drip line and plastic sheeting can be laid by machine	82
6.7	Sprinkling unit with rainwater tank as source	84
6.8	Rainwater concentration using plastic sheeting	86
6.9	Micro-catchments for tree planting	86
7.1	Terraces with plastic sheeting in Zhuanglang County	100
7.2	Contour planting for trees	110
7.3	Willow check dam	116
7.4	Newly built silt-retaining dam	117
7.5	Reclaimed land behind a dam	117
7.6	Zhang Yun's home	123
A1.1	The purification system	135
A1.2	An 80-year-old man enjoys the treated water	146

Boxes

2.1	Research results for runoff from different catchment types and rainfall events	22
3.1	Parameters of the solar cooker	34
4.1	Greenhouse development in Daping village	45
7.1	3-W model in the Jiuhuagou watershed	99
7.2	Tree and grass planting in the Qianjiagou watershed	111
7.3	Case study: The life-changing impact of watershed management	123

Acronyms and abbreviations

CNY	Chinese Yuan*
EVA	ethylene vinyl acetate
GRIWAC	Gansu Research Institute for Water Conservancy
HDPE	high density polyethylene
LORI	low-rate irrigation
NTU	nephelometric turbidity unit
OPC	Ordinary Portland Cement
PAFC	polyaluminium ferric chloride
PE	polyethylene
PVC	polyvinyl chloride
RCE	rainwater collection efficiency (runoff coefficient)
RWH	rainwater harvesting
WSE	water supply efficiency
WUE	water use efficiency

* Dollar equivalents have been given throughout the book at the exchange rate of the day: this is approximately 7 CNY = US$1. Where an earlier exchange rate has been used this is noted.

About the authors

Zhu Qiang is a retired Research Professor and former Director of the China Gansu Research Institute for Water Conservancy. He is a specialist in sustainable water resources management and rainwater harvesting. He was the Vice President of the International Rainwater Catchment System Association from 1999 to 2003.

Li Yuanhong is the Director and Research Professor of the China Gansu Research Institute for Water Conservancy. He is a specialist in rainwater harvesting and sedimentation management for reservoirs. He is a Council Member of the China Hydraulic Engineering Society and the General Secretary of the China Rainwater Utilization Special Committee.

John Gould is a rainwater harvesting consultant with over 30 years' experience undertaking research, training, assessments, and evaluations of rainwater harvesting projects in Africa and Asia, including China. He is a founder member of the International Rainwater Catchment Systems Association and was the Secretary-General from 1991 to 1995.

Preface

This book provides an insight into one of the great untold positive stories of the new century: how hundreds of millions of poor rural peasants in China's vast interior have greatly improved their lives, their environment, and their futures. At the centre of this story is a major rainwater harvesting programme. This provides a case study that outlines how this rapid social and economic transformation has been achieved in what has historically been one of China's poorest provinces, Gansu. The change that has taken place in many remote and water-scarce communities of this arid province in northwest China is nothing short of a revolution.

Our purpose is to provide the reader not only with an inspirational case study of a highly successful model of rural development, but with sufficient technical detail to allow for assessment of whether a particular technology or practice might be relevant for trialling elsewhere. While much has been written in Chinese about these technologies and sustainable water management practices, since 1995 there have only been a few articles in English on the topic, and access to much of the detailed information on these activities has been difficult for people outside China. This book, which is specifically focused on the recent developments in rainwater harvesting for domestic water supply and small-scale irrigation in the arid Loess Plateau region, is an attempt to open up the wealth of experience and knowledge built up in China over many years to a global audience through the medium of English.

The book is divided into four parts. Part I focuses on rainwater harvesting and domestic water supply and provides the background and context for all the innovations covered in this text. It also addresses the issue of water quality and the development of solar cookers. Part II addresses the application of rainwater harvesting for agriculture, including the development and widespread replication of highly productive low-cost greenhouses, with rainwater storage and micro-irrigation systems incorporated. The judicious use of stored rainwater, concentration of rainwater, and optimizing irrigation schedules to maximize the benefits of a limited supply and to supplement natural rainfall are also covered here. Part III examines rainwater harvesting and environmental management, addressing a number of aspects of small watershed management. Examples are terracing and other methods of soil erosion control, such as the establishment of trees and grass using micro-catchments and contour planting. Part IV addresses the challenges and future prospects for rainwater harvesting and some of the related sustainable technologies and management strategies that have developed in parallel, then closes with some conclusions.

While some of the chapters that follow may seem rather dry and technical, the impacts of the appropriate technologies and approaches described speak for themselves. In a world where in 2011 droughts and famines continued to ravage regions as diverse as the Horn of Africa and North Korea, the short-term responses in the form of emergency food relief actually continue to undermine the prospects for sustainable solutions by creating dependency. In similar poor and marginalized communities in the remote and arid regions of Gansu and other parts of China, however, the application of rainwater harvesting and other simple technologies have enabled people to 'drought-proof' their futures.

The concept of sustainable development is far from new, and at last there are signs that after decades of talking about it the world is starting to take this new paradigm more seriously. Since around 2007 there has been a growing realization among the public that the continuing pressures of climate change, environmental degradation, population growth, and resource depletion give us no choice but to adopt a more sustainable development trajectory in the future.

Through the establishment of the Millennium Development Goals in 2000 world leaders came together and set some ambitious targets for reducing poverty and improving access to basic healthcare, education, sanitation, and water supply (www.un.org/millenniumgoals). As 2015 steadily approaches it seems unlikely that more than a few countries will even come close to achieving all their goals and much remains to be done. While most of the world has struggled to turn grand plans for sustainable development into reality, in Gansu Province, China, a major demonstration of doing just this has been unfolding.

John Gould
November 2011

Acknowledgements

This book has been a joint collaboration between the co-authors based in China and New Zealand, with the support and assistance of many organizations and individuals in both countries and beyond. Through this acknowledgement we would like to thank all those who assisted us, and while it is not possible to mention everyone, we are particularly indebted to the following for their input.

In China we would like to thank: Ziyong Huang who kindly provided information and photos of solar cookers; Professor Xiaodong Guo for allowing us to use photos and other material from her presentation on greenhouses given to the Lanzhou International Rainwater Harvesting and Utilization Course for the Developing Countries in 2010; The Gansu Bureau of Soil and Water Conservation, and the Gansu Research Institute for Soil and Water Conservation for providing both photos and research reports on small watershed management. Our appreciation is also due to the Water Resources Institute of Inner Mongolia Autonomous Region for providing a number of photos; and the Water Resources Bureau of Dingxi, Huining, Qin'an, Jingning, and Zhuanglang counties of Gansu Province who assisted in collecting data on several of the case studies.

Special thanks also to the Gansu Research Institute for Water Conservancy for covering many of the expenses of writing this book, for inputs from several staff members and for providing research reports (see References section for further details) and the Innovative Team for the High Efficiency Use of Water Resources in Gansu for their support and for putting in an advance order for 300 copies of the book.

In Aotearoa, New Zealand, thanks to Althea Campbell, Liz Martyn, and the late Cynthia Guyan for proofreading the manuscript, and especially to Hugh Thorpe at the University of Canterbury, Department of Engineering, for his comprehensive technical review. Appreciation is due to the Christchurch Sister City Committee and to Ruth Fischer-Smith and staff at Christchurch City Council for their support and encouragement, as well as to Dave Adamson and other members of the New Zealand China Friendship Society.

Special thanks are also due for technical editing, review, and helpful suggestions from Professor Andrew Lo, President of the International Rainwater Catchment Systems Association (based in Taiwan), Professor Terry Thomas at the Development Technology Unit at the University of Warwick, UK, and to Karen Ekstrom in Uppsala, Sweden.

Finally, Clare Tawney and Toby Milner at Practical Action Publications in Rugby, UK, deserve our heartfelt appreciation for their patience, constructive comments, and guidance on improving the manuscript and overseeing its reproduction. While we have made every effort to ensure that all the material presented in this book is correct and error free, any mistakes that may have slipped through the exhaustive editing process are the responsibility of the authors alone.

Introduction

Since the early 1990s a major transformation has been taking place in numerous marginalized rural communities located on the Loess Plateau region in China's arid interior. Hundreds of thousands of families have been lifted out of poverty, had their health and diets improved, and the prospects for their children's futures enhanced. This has been achieved predominantly through the application of simple, affordable, and sustainable rainwater harvesting (RWH) and low-cost greenhouse and irrigation technologies. Rainwater harvesting is the small-scale collection, concentration, and storage of rainwater using structural measures, and its reuse for domestic, agricultural, or other productive purposes. These technologies have been used in conjunction with other appropriate technologies for improving domestic water supply, boosting crop production, and restoring small watersheds. In addition to improving livelihoods and the rural economy these developments are gradually transforming the whole environment, making once-barren land productive and covering once-desolate treeless mountains with forests and vegetation cover.

At the heart of this transformation has been the simple but highly significant upgrading of an ancient technology long used for storing rainwater: large, excavated, clay-lined underground tanks. The improvement of the traditional *shuijiao* (water cellar) technology has been achieved by applying modern science and the use of new materials such as cement. As a result, the capacity to harvest and effectively utilize rainwater has been greatly increased and the water and food security for thousands of communities in rural Gansu has been significantly enhanced. The integrated, holistic, and comprehensive nature of the RWH programme has been key to its rapid dissemination and adoption. This has been done with the guidance of the Gansu Research Institute for Water Conservancy (GRIWAC), which has painstakingly researched, field-tested, promoted, and disseminated RWH, along with soil and water conservation measures, since the late 1980s. The rapid replication of these successful innovations has been bolstered through a strong partnership between local government agencies and the community, allowing the programme to go from strength to strength.

The technologies and practices outlined are becoming increasingly commonplace in a large region cutting across China's arid interior and centred on the extensive Loess Plateau. This includes most or parts of the provinces of Gansu, Shaanxi, Shanxi, Ningxia, and Inner Mongolia. In total the Loess Plateau covers over 510,000 km^2 and despite the inhospitable nature of much of this arid region, it is home to around 40 million people. In this book, the focus is primarily on the impact of these innovations in Gansu Province, as here the changes have been especially dramatic and many of the components of the transformation well documented. Gansu Province has long been considered one of the poorest in China. For centuries the underlying causes of poverty in Gansu and marginalized provinces on the Loess Plateau have been water scarcity, drought, and land degradation. These are common challenges faced by rural communities in arid and semi-arid regions across the developing world and especially in many of the drier parts of Africa, where poverty, drought, and periodic food shortages remain a perennial problem.

It is, however, important to acknowledge that while many of the technologies and approaches outlined have been very successful in China, they were developed to meet the needs of the communities living in the specific geographic, climatic, cultural, and social context of Gansu Province. Nevertheless, while only some of these innovations may be appropriate elsewhere and significant local adaptation will be needed to encourage replication beyond China, the principles underlying the success of this revolution in sustainable rural development are universal.

These core principles include the need in any programme to:

- build on local knowledge and traditional wisdom;
- put communities at the heart of every stage of any new initiative, whether introducing a technology or encouraging a change in a traditional farming practice;
- field-test and pilot any new development with the future users and listen to their experiences and suggested improvements;
- ensure that technologies or changes in practice are affordable and viable – or that mechanisms can be put in place to make them so, for example loans or subsidies.

Some of the technologies and practices being used in this quiet revolution towards sustainably managing natural resources may be transitional in nature. For example, the use of individual household solar cookers may become obsolete once communities are connected to the electricity grid. Others, however, are likely to be needed and used for the long term, such as the terracing works and the ecological restoration activities described in Chapter 7. These not only offer the prospect of increasing agricultural activity in the region, but as both soil and vegetation are huge potential reservoirs for atmospheric carbon, they offer the potential to assist in countering the rising levels of atmospheric carbon dioxide over the coming century, thus helping to slow and ameliorate the impact of climate change.

PART I
Rainwater harvesting and domestic rural water supply

CHAPTER 1
Development of rainwater harvesting in rural Gansu

Gansu climate and water resources

Gansu Province is located in northwest China (Figure 1.1) and has a population of 26 million. It is one of the driest provinces in the country. The mean annual precipitation is only 306 mm while the potential evaporation ranges from 1500 to 2500 mm. Water scarcity has obstructed social and economic development and all efforts to improve people's lives. Table 1.1 shows the precipitation and water resources in the world, China, and Gansu Province. This chronic water shortage has been one of the main causes of low agricultural production and people's poverty in the area.

The situation is even worse on the Loess Plateau located in the central and eastern parts of the province. The extensive Loess Plateau and hills are crisscrossed by numerous gullies and ravines. Most of the rainfall is absorbed by the highly permeable loess[1] soil, with a depth ranging to 300 m. The geomorphology of the whole region is characterized by the Loess Plateau, which was deposited by wind and river action over a very long period of time. With a mean annual runoff coefficient of only 0.08 to 0.1, only a small part of the rainfall forms river flow. In most areas, the loess layer lies on impervious Tertiary rock, so groundwater is also very rare. With total annual available water resources of only 260 m³ per person, the conditions on the Loess Plateau are at the lower limit for sustaining human existence.

While actual precipitation ranges between 200 and 500 mm/year in most cultivated areas, 55 to 60 per cent is concentrated in the late summer period from July to September, with only less than a quarter occurring during the critical spring period for crop growth.

Table 1.1 The precipitation and water resources in the world, China, and Gansu

Region	Annual precipitation, km³	Annual water resources, km³	Water resources per capita, m³	Waterresources/ precipitation
World	108,000	47,000	7300	0.406
China	6,190	2,800	2100	0.452
Gansu	129.7	27.2	1050	0.211

Note: Water resources denotes surface water plus groundwater minus overlap of the two water sources.
Source: Data for world annual precipitation and water resources are from Seckler D. et al. (1998) *World Water Demand and Supply 1990–2025: Scenarios and Issues*, IWMI; for China and Gansu: Qian Zhengying (ed.) (1991) *Water Conservancy in China*; population data: UNESCO (2006) *Water, a shared responsibility: UN World Development Report 2*, available from <www.unesco.org/water/wwap>

6 EVERY LAST DROP

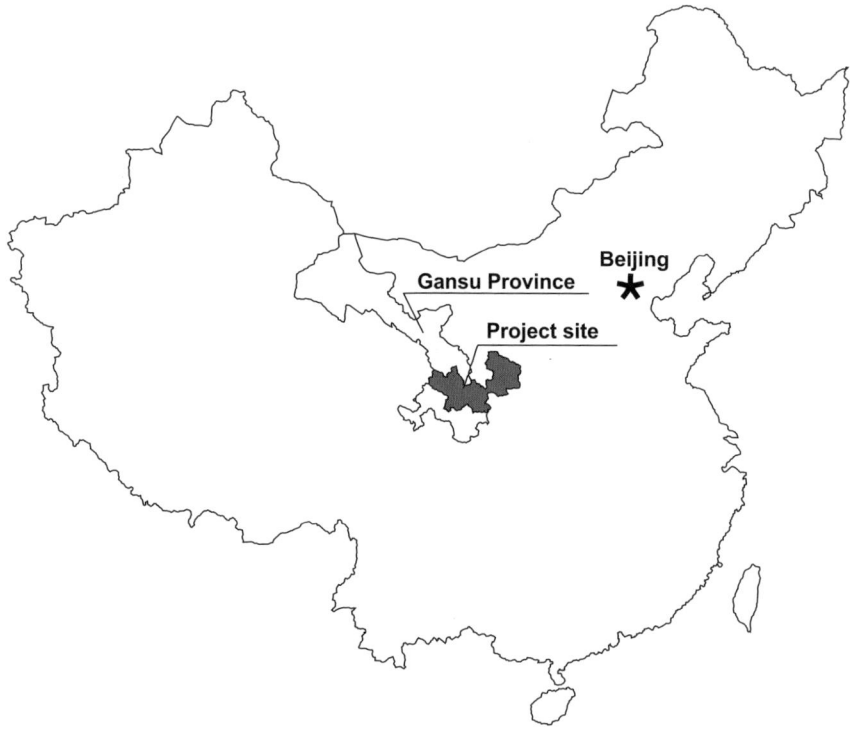

Figure 1.1 Map of China showing Gansu Province
Source: Adapted from http://nfgis.nsdi.gov.cn

Droughts occur frequently and historical records over the past 1,400 years since the Sui Dynasty indicate that droughts have occurred in 634 years or almost one in two years. However, in the 41 years from 1966 to 2007 there were 36 droughts, close to nine in every ten years! This problem of growing water scarcity has had serious consequences for both the people living in the area and the environment:

- Before 1995, around 3 million people in Gansu had no access to safe drinking water. During dry periods the government trucked drinking water from distant sources such as small streams, as owing to the geological conditions most of the groundwater is saline with a high mineral content. At those times and in places where trucked water was not available, people took several hours each day to carry water from remote sources located in deep valleys. Surveys at the time indicated an average labour requirement for a family of five of 70 days annually. At that time it was also common for both humans and animals to share open pond water sources, with all the associated health risks that implies. Another source was water harvested from traditional *Shuijiao* rainwater systems (see over), which was both turbid and dirty.

DEVELOPMENT OF RAINWATER HARVESTING

- The water shortage meant that subsistence farming was the norm as without a reliable water supply, farmers could not grow cash crops and as a result they had very low incomes. The yearly net income per capita was only US$60–100 and about 4 million people lived at or below the poverty line. Before the 1990s most people in the mountainous areas could never produce 'fine' vegetables such as peppers, tomatoes, cucumbers, and egg plants.
- Agriculture depended on unreliable rains. The mean grain yield was only 1.8 t/ha (284 kg per person). Owing to the low productivity, farmers were forced to cultivate as much land as possible, even on steep slopes. This situation caused increasing land degradation through soil erosion with losses of 5,000–10,000 t/km², equivalent to a layer of topsoil of 0.5–1 cm, being stripped off the surface each year.

Due to the scattered nature of the population, the difficult topography, and the high economic cost of any form of water diversion scheme, addressing the water shortage in the region was very difficult. The only viable potential water resource in the area was the direct use of rainwater. Despite the relatively low rainfall in the region, the total amount of rainwater falling on the whole area of almost 150,000 km² is equivalent to 54 billion m³ per annum, much more than the yearly runoff of the Yellow River at Lanzhou, capital of Gansu. Most of this water, about three-fifths, is lost through evaporation, a fifth forms river runoff, and only the remaining fifth is directly used by crops.

Utilization of rainwater in such a dry region poses many challenges as not only is the total annual rainfall low but it is unevenly distributed throughout the year, being concentrated between July and September. Droughts, sometimes

Photo 1.1 Soil erosion and cultivation on the steep land are both common features of the Loess Plateau in Gansu Province

lasting years, are not uncommon. For rainwater use to be viable it is therefore imperative to concentrate the rainwater runoff and store it for future use.

The harvesting of rainwater for future use is not a new phenomenon on the Loess Plateau but builds on a very long tradition. For centuries many families have excavated household water cellars known locally as *Shuijiao*. These bottle-shaped underground tanks lined with clay stored runoff from mainly natural soil surfaces. Most families would have relied on about two water cellars, which typically had volumes ranging from 20 m^3 to as much as 70 m^3. Due to the lack of inflow, seepage, the low runoff coefficient of the soil, and in many cases the insufficient capacities of the *Shuijiao*, these could generally not supply all the domestic household water needs and certainly not enough to support any supplementary irrigation. Such traditional systems also required a lot of maintenance, with the need to constantly paste straw and clay to the walls and base to reduce seepage losses. The red clay used for this was found in a layer beneath the loess and provided a reasonably watertight lining. Leakages, however, sometimes eroded the loess soils supporting the *Shuijiao*, leading to collapse. Another problem with the traditional technology was that while light rains produced little runoff from the untreated catchments, heavy rains resulted in erosion and turbid water of low quality filling the tanks. In the late 1980s many rural households in the region still depended on the traditional *Shuijiao* and most of them suffered perennial water shortages.

In order to improve this situation, a research and demonstration project was initiated by GRIWAC from 1988 to 1992, with the support of the Gansu provincial government and the Gansu Bureau of Water Resources. The GRIWAC team set up 28 testing plots and used eight different materials for constructing catchments. The rainwater collection efficiency (RCE) was tested and the RCE versus rain characteristics (rainfall amount and intensity) for all eight materials calculated. The findings of this work revealed that the natural soil surface catchments traditionally used had an RCE of only 0.08. The RCE for traditional mud straw roofs in rural areas was also very low. However, the research showed that by upgrading the roof using tiles, the RCE could be raised significantly, to 0.75 for cement tiles or 0.45 for clay tiles. The figure for a concrete lined courtyard surface was found to be as high as 0.8 (up to a hundredfold increase). With these research findings, the GRIWAC team recommended the widespread use of these methods of harvesting rainwater as a model for rural domestic water supply. By using a tiled roof and concrete-lined courtyard as the catchment and an appropriately sized cement-lined underground water cellar for rainwater storage they calculated that basic water needs could be met in 9 years out of 10. A three-year pilot demonstration followed which showed that this system was highly effective for solving the domestic water problems faced by the rural population.

The GRIWAC team also demonstrated that supplementary irrigation using the same basic RWH approach was feasible. Simplified low-cost greenhouses built using plastic sheeting as the catchment could provide enough runoff to meet the irrigation water demand for a vegetable crop for one harvest.

Based on the successful results of this research and demonstration project, the Gansu provincial government decided to initiate the 1-2-1 Rainwater Catchment Project,[2] the first large-scale RWH project in China, in the summer of 1995. The occurrence of a 1-in-60-year drought in the Loess Plateau region of Gansu that year created enormous hardships for the rural population of the province and clearly prompted the decision to provide urgent and significant support. The 1-2-1 Project was very successful and solved the problem of domestic water shortages for 1.3 million people.

The first trial brought tremendous change to numerous impoverished villages in Gansu and also demonstrated to the authorities and local people the great potential offered by RWH. Since 1996, RWH in Gansu has been extended from mainly supply for domestic use to also enhancing crop production, with its coverage now extended to the whole area of rain-fed agriculture.

More than 20 years of trialling different practices has proved that RWH not only can meet the basic water need of rural households in Gansu but can also supply water for supplementary irrigation, with a resulting average increase in crop yield of 40 per cent. The experience has shown that RWH can provide an alternative way for communities in dry mountainous areas on the Loess Plateau to get water. The shortage of water has been the root cause of the impoverished life that many rural inhabitants in the region have led for centuries. With water in their tanks, farmers can rearrange their production patterns according to market needs, thus greatly enhancing their income and alleviating poverty. Improvement of land productivity through supplementary irrigation has helped the farmers also accept state-supported land conversion

Photo 1.2 Rainwater harvesting using tiles and concrete paving, for domestic water use

Photo 1.3 Rainwater harvesting for supplemental irrigation

Photo 1.4 Rainwater harvesting systems for ecological conservation

programmes, encouraging them to shift from low-yield crop production on steep hill slopes to planting trees and grass, thus promoting environmental conservation through reduced soil erosion.

RWH is a decentralized solution for water management that is particularly suitable for populations scattered in mountainous areas. Since the technology is simple and builds on traditional tried and tested techniques it is well accepted by the farmers. Since implementation is predominantly done by the householders with limited government support, the water users feel ownership of the systems that then legally belong to them. Householders therefore are highly motivated to participate in every stage of project implementation, from planning, design, and construction to operation and maintenance. Unlike large-scale water development schemes that often create significant impacts on ecological systems and can be socially divisive, household-level rainwater harvesting projects are both environmentally friendly and attract significant community support.

The harvesting and use of rainwater as described above has now become part of an integrated approach for sustainable development in the mountainous areas of Gansu and neighbouring semi-arid regions. This general approach has also been replicated in the sub-humid and humid areas in 15 provinces (autonomous regions and municipalities) in China. Many large-scale projects such as the 1-1-2 Water-Saving Irrigation Project[3] in Inner Mongolia, the Water Cellar Irrigation Project in Ningxia, the Thirsting for Water Project in Guizhou, the Sweet Dew Project in Sichuan, etc., have followed the Gansu experience and have brought about significant benefits to millions of rural people. According to an investigation in 2007, the number of rainwater tanks (and small rainwater storage reservoirs and ponds) constructed in these provinces amounted to over 10 million units, with total storage capacity of 4.6 billion cubic metres. These rainwater systems have provided domestic water supply to 22 million people and supplementary irrigation for 2.8 million hectares of rain-fed farmland. RWH has thus become an important component of water resource development in China, which is an integration of large, medium, small, and mini-sized schemes.

Rainwater storage tanks

Water is highly valued in the Loess Plateau of Gansu Province, in fact in the whole of northwest China. In the past, the best gift one could give when visiting a friend or relative was a vessel of good quality water. In their long history of fighting against thirst, the rural people in the middle and eastern parts of Gansu have a long tradition of using rainwater as their main source for domestic supply. The underground tank used for rainwater storage, *Shuijiao*, is composed of two Chinese words, *Shui* meaning water and *Jiao* meaning cellar. The number of *Shuijiao* owned by the family has also become a status symbol. When a young woman is to marry, her parents might raise the question with her prospective in-laws: 'How many water cellars do you have?', the number of water cellars being a key indicator of the future quality of life for the new couple. People have come to regard the *Shuijiao* as just as important as the new home.

Traditionally, the cellar was an underground room where rural people stored their vegetables and other foods. Since the winter is very cold in Gansu, with temperatures dropping to -20°C, vegetables cannot grow. To have enough vegetables (mainly cabbage and potato) for the long winter, people dig a cellar 2 m underground for storing food harvested in the autumn. The temperature underground remains in a suitable range for the storage of potatoes and cabbage until the next harvest. The water cellar developed from the cellar for storing food and other items but was different in structure. The reason people adopted underground tanks to store rainwater was that in the past they mainly collected the runoff from natural slopes, roads, and courtyards: most of the roofs were covered with straw and mud, and produced only limited runoff of poor quality. Other reasons also included:

- An underground tank can avoid or minimize evaporative loss that can amount to more than 1 m annually in this semi-arid climate.
- With favourable loess soil conditions and structure, the underground tank is cheaper to construct than an above-ground structure.
- Water stored in an underground tank keeps cool. If the water table is located 1.5 m under the surface then the water temperature is close to the mean annual air temperature of around 8°C in Gansu. This low temperature is favourable for storage of good quality water.
- On the Loess Plateau of Gansu, where winter temperatures may drop to -20°C, storing water underground prevents it from freezing.
- An underground structure can also save land. In Guizhou Province, for example, where land is scarce, farmers usually make use of the land on top of the water cellar for planting maize, vegetables, etc.

Traditional *Shuijiao*

The traditional *Shuijiao* constructed in the region is bottle shaped, see Figure 1.2. The ratio of the depth to the maximum diameter is usually 1.5–2. Such

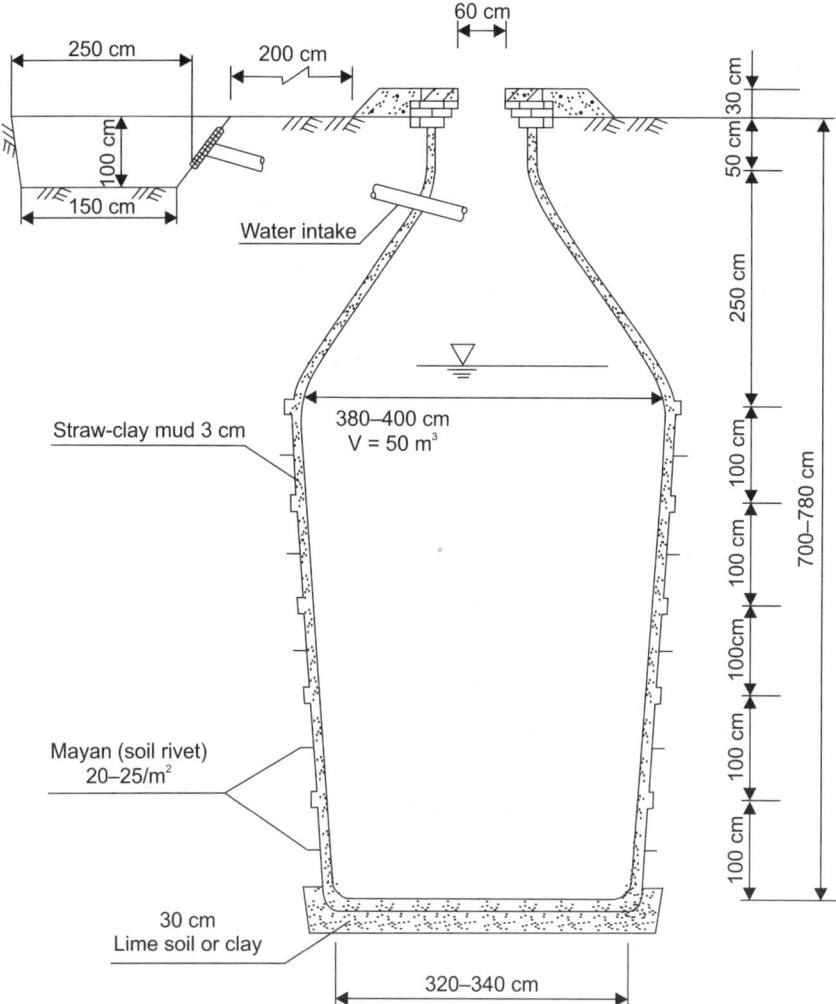

Figure 1.2 Traditional *Shuijiao* water cellar using clay-mud paste for seepage control

traditional water cellars have three parts: the tank opening, the dry part, and the water part. The tank opening is located above the ground. It has a diameter of 0.8–1 m, a height of 0.6–0.8 m, and is built of brick or stone. The dry part of the tank is usually not for storing water. It is the shape of a hopper and is the transition from the largest inside diameter of 3–4 m to the opening diameter of less than 1 m. The inverted slope is about 1 to 0.3–0.5 (vertical to horizontal), depending on the soil strength that ensures stability. The wall of the lowest part that stores the water is vertical or with a slight positive slope. The bottom is flat or in the shape of an inverted arch.

Building such water cellars cost effectively, i.e. with minimum inputs, has been very important to the local people. During their long history, people have developed an ingenious way of constructing the water cellar in the loess soil. The unique soil structure of loess has the capability of standing vertically up to 10 m or more in dry soil, without support. The local people have made full use of this property to build the *Shuijiao* using very few construction materials, only stones, gravel, and sand. However, if the soil is moistened by infiltration following a storm it may slump and cause damage to any structure built either on or in it.

The essential element for successful water cellar construction is to ensure that no infiltration of tank water into the soil occurs. The material used for preventing seepage should be locally available and affordable. Building techniques with this kind of material also have to be simple enough to be acceptable to local people.

In ancient times, before cement was available, clay mud was used as the sealant for the water cellars. The walls and base were plastered with a red clay mud mixed with straw. It was of vital importance to ensure the quality of the clay-mud layer as this had to be completely impermeable and free of any cracks or fissures during plastering. This layer also needed a close and firm connection with the surrounding soil to prevent collapse. The construction procedure can be summarized as follows:

1. The red clay should first be broken into fine particles by using a hammer and then passed through a 5 mm screen.
2. The crushed clay is immersed in a shallow pond with an appropriate amount of water, and after the clay has fully absorbed the moisture, straw is mixed into it. People trample on the clay-straw mud to make the mixture uniform.
3. The red clay-straw mix is kneaded and shaped into malleable bars with diameters of about 5–6 cm.
4. The fully excavated *Shuijiao* is next ready for plastering. Small holes shaped like truncated cones, penetrating about 10 cm into the wall with outside diameters of 5–6 cm and inner diameters of 8–10 cm, are made on the wall, not by digging but by hammering a rod into the loess soil. The hole is used for inserting the clay 'rivet', which has the local name of *Mayan*. There are about 20–25 rivet holes per square metre.
5. The malleable mud bar is pressed firmly into the rivet holes to create a good bond with the native loess soil. These rivets are then linked by adding more of the straw-mud mixture to form an integrated clay layer with a thickness of about 3 cm covering the whole wall.
6. A wood hammer is used to compact the clay-mud layer and make it dense and firm.
7. An additional 1 cm thick layer of clay mud, free of straw, is plastered on the first layer.

8. After plastering is finished, flax (or a local plant with a fibrous stem) is soaked in water and used to brush the surface to enhance anti-seepage capability. Further compaction using a hammer continues daily. The whole operation takes around 10–15 days.

The excavation technique of the cellars has also evolved through long experience. To make a thin-walled water cellar, local people traditionally use the following steps:

1. The cellar is centred on the ground.
2. Excavation of the cellar is started within a 60–100 cm diameter circle, the soil being removed by shovel.
3. When the hole becomes too deep for a man to throw the earth out of it, a wooden tripod is erected above the hole with a wheel fastened to it. The earth is then lifted out with a bucket using man or animal power.
4. The excavation takes place from the centre outwards. A plumb-line is used to keep the excavation centred. The last 3–4 cm of soil is left uncut and is compacted using a wooden hammer to increase the density of the soil on which the clay mud will be pasted.

CHAPTER 2
Upgraded water cellar designs

During the research and demonstration project carried out by GRIWAC, three basic designs of water cellar were developed, each suited to a different soil texture and density, namely: a thin-walled water cellar, a water cellar with a concrete dome support, and a cylindrical water cellar.

Thin-walled water cellar

The thin-walled water cellar is suitable for firm soil (density higher than 1.4 t/m^3 and clay content over 15 per cent) in which a vertical cut with a height of more than 10 m can be kept stable when the soil is dry. The whole wall of the cellar can be kept steady just by plastering the wall (both dry and wet parts). The key issue is ensuring the thin layer is watertight and fully integrated with the surrounding soil. Traditionally, the material used for preventing seepage was clay mud. However, construction using the clay-mud layer was very complicated and labour intensive, and quality assurance was difficult. Damage and even collapse of traditional cellars was common owing to infiltration of water through the cracks in the mud layer into the soil, causing subsidence. GRIWAC recommended using cement mortar as a substitute for the clay mud. Cement mortar is easier to plaster than clay. A 3 cm-thick wall is acheived by plastering the mortar three times, applying a 1 cm layer each time. The mortar should be grade M10 or higher, i.e. having a compression strength of 10 MPa (or 100 kg/cm^2). To ensure this standard, the proportion of cement, sand, and water is critical and should be determined through careful testing. Usually the proportion depends on the properties of the local sand (strength, texture, shape, etc.) and the grade of cement. To give a rough idea, the M10 grade mortar can be produced with an Ordinary Portland Cement (OPC, Grade 32.5) to sand to water ratio of 1:4.0:0.7 by weight. Each cubic metre of cement mortar requires 387 kg of cement (Grade 32.5), 1,560 kg of sand, and 275 kg of water. If cement OPC Grade 42.5 is used, then the proportion will be 1:4.4:0.77. For each cubic metre of mortar, 357 kg of Grade 42.5 cement, 1,586 kg of sand, and 275 kg of water are used. After plastering of the cement mortar is finished, a pure cement grout (cement to water ratio of 1:1–1.5) should be washed on the mortar surface two to three times. After completion, curing of cement mortar is done by moistening the surface with water once every day for at least 10 days.

If the natural soil at the bottom is not dense enough (less than 1.4 t/m^3) a 30 cm-deep layer needs first to be compacted in layers of 15 cm maximum thickness. The bottom of the tank is the part subjected to the highest water pressure. For additional strengthening, a layer of compacted lime soil (a mixture of lime and soil with proportion of lime to soil of about three to seven by volume) with a thickness of 20–30 cm can be added on top of the compacted natural soil before the base is plastered with cement. On top of the lime soil, a layer of cement mortar the same thickness as the wall is plastered, or alternatively 10 cm of concrete is cast. The cement mortar has a grade of M10 and the concrete has a grade of C15. The section of the thin-walled water cellar is similar to that in Figure 1.2.

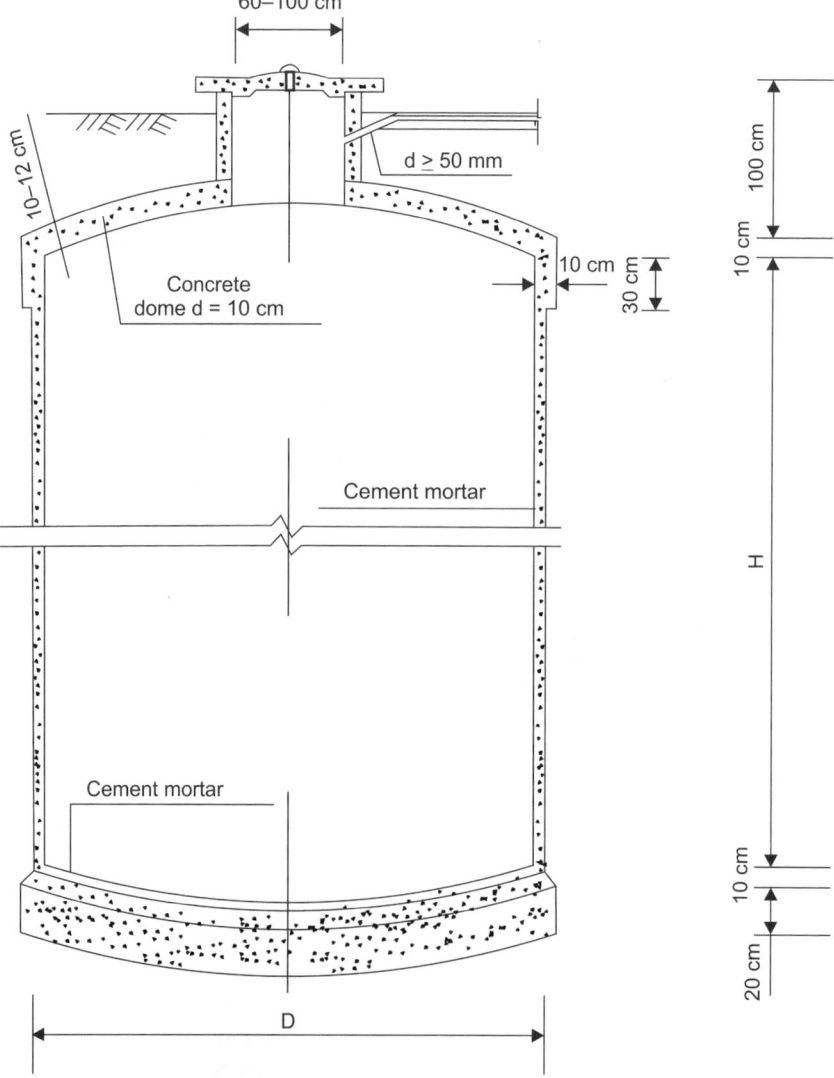

Figure 2.1 The water cellar with top semi-dome structure

Concrete domed water cellar

The concrete domed water cellar is constructed when the soil is not so firm and the top of the cellar needs greater support. Instead of a cement mortar lining, a dome structure of plain concrete is constructed for the top of the cellar. The semi-dome can also be built with bricks if the builders are skilled enough (this method is described below). The dome is 10 cm thick and uses grade C15 concrete. The wall structure is the same as for the thin-walled cellar described above.

For casting the concrete semi-dome structure at the top of the cellar, local people have developed the following cost-effective method (Figure 2.2):

1. Excavation of the cellar starts with the outer diameter of the semi-dome. Digging is from the ground surface and shaped to the profile of the lower surface of the semi-dome (Figure 2.2a). An allowance of 2–3 cm is left for compaction of the soil surface on which the dome will be cast.
2. Casting of the concrete takes place using the earth as the mould (Figure 2.2b) and a hole for the cellar opening should be left free.
3. After one week of hardening of the concrete, the excavation is restarted from the opening of the cellar (Figure 2.2c and 2.2d). This is the excavation procedure for the thin-walled cellar (i.e. the hollowing-out process). An allowance of 2–3 cm is left when cutting both the wall and bottom, which are then compacted by hammering to the correct dimension. The aim of this is to compact the soil behind the walls making it denser.
4. After finishing excavation, the plastering of the wall with cement mortar is carried out starting at the top (Figure 2.2e).
5. Then the base concrete slab (or inverted dome floor) is cast on the compacted soil or to strengthen the anti-seepage performance on a layer of compacted lime soil.
6. Finally, the opening of the cellar is installed, using a prefabricated reinforced concrete pipe (Figure 2.2f).

This hollowing-out method of excavation can significantly reduce the workload and avoids having to use a wooden mould and framework that are expensive and time consuming.

When bricks are used for the semi-dome structure, conventional practice requires the erecting a supporting framework using expensive timber or steel rods. However, skilled local labourers can build this kind of structure without any support using only a simple platform for standing on, as shown in Photo 2.1. As the brick layers are completed, cement mortar is plastered on the outer wall of the dome. When the dome reaches the top, the opening of the cellar can be built with a concrete ring or with bricks.

18 EVERY LAST DROP

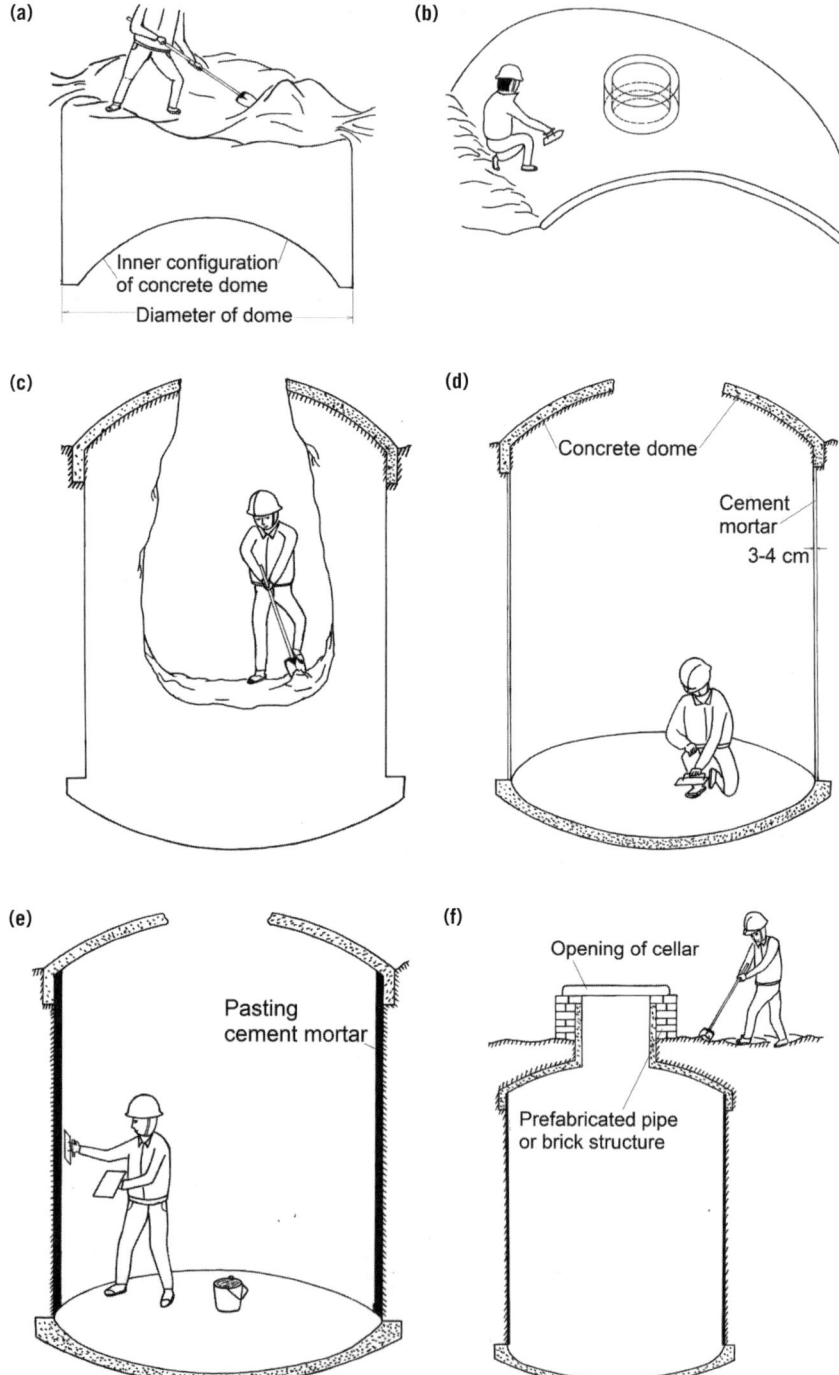

Figure 2.2 Construction procedure for building water cellar with top dome structure

UPGRADED WATER CELLAR DESIGNS   19

Photo 2.1 A simple platform enables the building of a brick semi-dome for the top of a water cellar

Cylindrical water cellar

The cylindrical water cellar is an appropriate design when the soil is weak and sandy, and the tank structure needs to be strengthened. The cylindrical wall is built of C15 plain concrete cast in situ (see Figure 2.3).

The thickness of concrete is usually 10–12 cm. The cylinder can also be built with prefabricated concrete blocks or with bricks. In these cases, a layer of cement mortar should be plastered on the masonry to enhance the performance of seepage control. Brick-built cylindrical cellars are suitable for mild climates. For

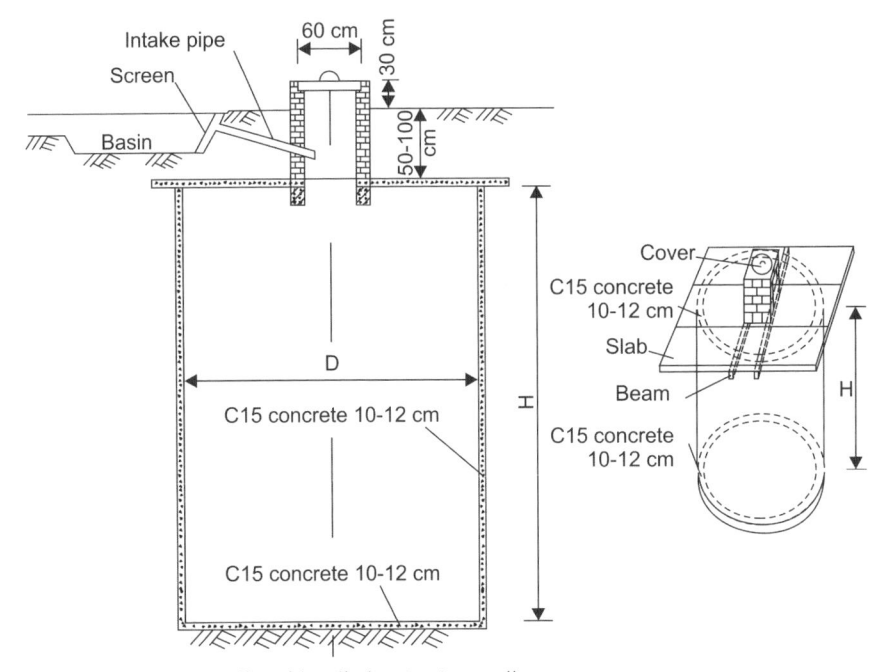

Figure 2.3 The water cellar with cylinder-structure wall

20 EVERY LAST DROP

the top of the cellar, three types of structure can be used, namely, the brick-built semi-dome structure, the in-situ-cast plain concrete semi-dome structure, and the reinforced concrete slab or reinforced concrete slab-beam[4] structure. The reinforced concrete-slab structure can be used when the diameter of the cellar is 3 m or less, as this can be put directly on the edge of the concrete cylinder. If the diameter exceeds 3 m, two beams should be used to support the slab (Figure 2.3). The base comprises a concrete slab cast in situ with a thickness of 10–12 cm.

Cost estimates for different kinds of water cellar are shown in Table 2.1 and Figure 2.4.

It can be seen that the cost of a water cellar with a thin cement mortar plaster structure is much lower than the concrete semi-dome top structure and the reinforced concrete cylinder structure. Where possible the thin-walled water cellar is preferable because it is the cheapest. In addition, the bigger the volume of the cellar the lower unit cost (cost per volume). Therefore, for the same storage volume one large cellar is always cheaper than two small.

Table 2.1 Unit cost of different upgraded *Shuijiao* (CNY)[5]

Type of water cellar	Unit cost for water cellar with different volume (CNY/m³)			
	15 m³	20 m³	25 m³	30 m³
I	52.4	43.1	37.4	35.6
II	61.6	55.6	53.9	50.3
III	86.8	75.8	69.3	64.9

Type I: thin-walled water cellar   Type II: concrete domed water cellar   Type III: cylindrical water cellar

Note: 8.3 CNY = US$1

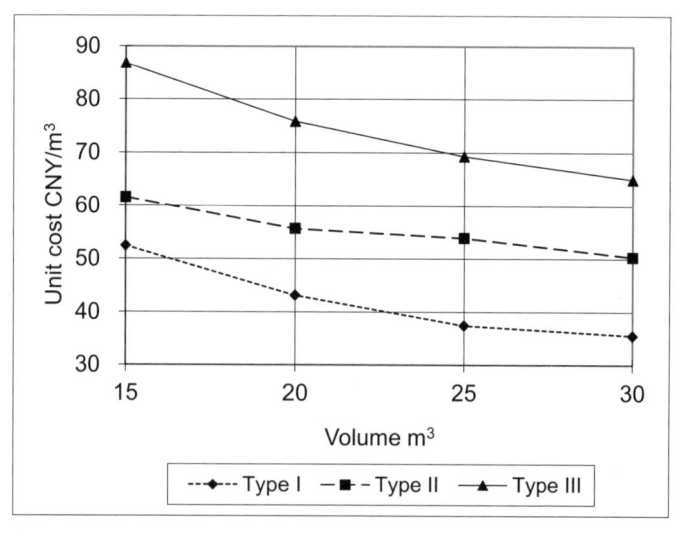

Figure 2.4 Cost of *Shuijiao* (unit cost versus volume)

Photo 2.2 Water delivery facility using a hand pump and a bucket and rope

After seven years of demonstration and small-scale extension, a domestic RWH system was developed as a model for replication. It consisted of three components: the rainwater collection surface composed of the tiled roof and concrete-lined collection surface, the water cellar as storage, and the water delivery system. The delivery system varied depending on household preference. About 53 per cent of the households used a simple bucket and rope, 18 per cent a hand pump, and 29 per cent a submerged electric pump. The preference for using a rope and bucket was partly due to cost but also this was considered the simplest and most reliable method, despite the risk of contamination. The hand pump costs about $25 and is affordable for most households; however, it is not considered very convenient because the pump has to be primed each time after stopping use for a certain period.

Rainwater harvesting system design procedure and development

During the research and demonstration project on RWH carried out from 1988 to 1992, a key finding was the need to raise the runoff coefficient or Rainwater Collection Efficiency (RCE) of the systems. In the past, great attention had been given to the development of storage but little attention to the effectiveness of the catchment. Most traditional catchments comprised natural earth slopes having a very low RCE in the range of 0.08–0.1. The runoff from these natural rainwater collection surfaces was often insufficient to fill the water cellars. In 1988, on average 2.2 water cellars were owned by each family when GRIWAC started their research and demonstration project. Many of these, however, were only partly filled or even empty, and this clearly had not encouraged and motivated farmers to build more water cellars. Through the research and development work undertaken by GRIWAC, the RCE of catchments has been greatly enhanced, and for domestic RWH systems now being constructed the RCE is 0.7–0.8, which is almost a tenfold improvement.

After the research work was completed, suitable catchments for different types of water use were recommended. For the RWH systems for domestic water supply, tiled roofs and concrete-lined courtyards were promoted. Where the original roof was made of straw mud, replacement with a tiled roof was

22 EVERY LAST DROP

encouraged. Several types of tiled roof exist in the area. In Table 2.2 it can be seen that the cement tile roof has an RCE of 62–75 per cent, which is up to twice as much as that for factory-made clay tiles. Since the price of the tiles is about the same, it makes sense to use the cement ones. For systems where RWH was developed for irrigation, surfaces with low permeability such as paved highways, country roads, threshing floors, sports grounds, etc. were recommended. Natural slopes can also be used as catchments for irrigation purposes. Due to their larger areas, a lower RCE is feasible and since the water is not for human consumption, considerations of water quality are less important.

One very practical result of the research was the development of a design procedure for sizing the required catchment surface area and matching it to the available storage. Provided the storage volume is sufficient, the catchment area required for collecting enough water to meet a given demand can be calculated using the following equation:

$$A = \frac{1{,}000 \times W}{R_d \times RCE} \tag{2.1}$$

where A is the area of catchment in square metres, W is the water supply for domestic use in the whole year in cubic metres, RCE is the rainwater collection efficiency, and R_d is the design yearly rainfall in millimetres. The design yearly rainfall denotes the yearly rainfall having an exceedance of 90 per cent in a long series of yearly rainfall (i.e. a 90 per cent chance that this rainfall amount will be exceeded in any given year and the 90 per cent relates to the reliability of the rural domestic water supply). Tables for calculating the

---

Box 2.1 Research results for runoff from different catchment types and rainfall events

GRIWAC spent four years testing the RCE for different types of catchment. In total 28 plots for testing the RCE of eight catchment types were set up and 421 rain tests, including natural rain and artificial rain (by sprinkling in the field and lab) were conducted. The rainfall amount and intensity[6] in each rain event as well as the moisture content on the collection surface were measured. The RCE versus the rainfall amount and intensity for the eight types of catchment were analysed and regression equations for calculating the RCE were computed. The summarized results giving the RCE values for the various catchments studied are listed in Table 2.2.

Here the yearly mean RCE refers to the average of RCE in one year. It will vary depending on the rainfall conditions of the area and from year to year. It shows that catchments of the same material have different RCE in areas with different mean annual rainfall. In any one area the RCE is higher in the wetter years than in the drier years. This is because in wetter years heavy rain events happen more often than in drier years, and the RCE of a heavy rain event is higher than in a light rain event. Since the Loess Plateau of Gansu is a semi-arid area, most of the rain falls in lighter events. In humid areas there is a much higher percentage of heavy rains and the catchment surface is usually wetter than in Gansu, so the RCE is much higher. In the *China National Technical Code of Practice for Rainwater Collection, Storage and Utilization* (China Ministry of Water Resources, 2001), the recommended RCE values for different kinds of catchment in the semi-arid, sub-humid, and humid areas are provided.

Table 2.2 Yearly mean RCE of different types of catchment (%)

Annual rainfall (mm)	Concrete	Cement soil	Buried plastic film	Compacted loess	Cement tiles	Clay tiles made in factory	Clay tiles made in workshop
400–500	76–80	42–53	36–46	19–25	69–75	39–50	31–41
300–400	75–80	40–52	34–46	17–26	67–75	37–49	29–40
200–300	73–78	33–47	28–41	13–20	62–71	30–41	24–34

catchment area are provided for farmers with primary education in order for them to determine the catchment area of their own RWH system. These tables can be found in the technical codes of the China Ministry of Water Resources (2001) and are reproduced in Zhu et al. (2007).

Clay tiles can absorb a lot of water compared with cement tiles or concrete slabs and hence have a lower RCE.

Since the cost of storage is about 50–70 per cent of the whole system, it is important to be able to estimate the required volume as accurately as possible. In addition, because RWH systems are implemented by farmers, the method for determining the volume should be simple and easy to understand. To design the storage volume, two methods can be used: the simplified method (volume coefficient method) and the simulation method.

The simulation method is described in detail in Zhu et al. (2007), in Volume I, Part 2. This method needs to be applied by water technicians at the local level. The simplified method, however, can be easily understood by the famers with primary school mathematics skills. It gives the storage volume as a fraction of the water amount supplied by the RWH system in the whole year as shown in the following equation:

$$V = K \times Ws \qquad (2.2)$$

where V is the volume in cubic metres, Ws is the water supplied in the whole year in cubic metres, and K is the volume coefficient used to calculate tank volume. The coefficient K is shown in Table 2.3 based on a hydrological calculation using a long series of rainfall data from rain gauges located on the Loess Plateau in Gansu and in the humid area of China. K is obtained by dividing the tank volume with the annual rainwater inflow. Since it is relatively easy to estimate the amount of runoff from the catchment, this simplified formula (2.2) makes the determination of the tank volume for different purposes straightforward.

Table 2.3 The volume coefficient K in Equation 2.2

Domestic supply	0.55–0.6
Supplemental irrigation for crop field	0.85–0.9
Greenhouse with irrigation all year round	0.55–0.6

Note: In other climates than that of Gansu, K will have to be calculated by a local water specialist. Where rainfall is less seasonal, recommended K values for domestic supply may be lower than for Gansu.

Table 2.4 Dimensions of circular water cellars

Volume (m³)	Water depth (m)	Max. inner diameter (m)	Thin-walled water cellar Semi-dome	Thin-walled water cellar Total	Concrete domed top Semi-dome	Concrete domed top Total	Cylindrical water cellar Total
20	3.7	2.6	1.0	5.0	0.9	5.1	4.7
30	4.1	3.0	1.2	5.5	1.0	5.3	5.1
40	4.2	3.4	1.4	5.7	1.0	5.4	5.2
50	4.4	3.7	1.5	6.0	1.1	5.7	5.4
60	4.5	4.0	1.6	6.2	1.2	5.9	5.5

Once the total storage volume required has been established, the volume for each tank can be determined by dividing the total volume by the number of tanks. It can be seen from Figure 2.4 that the unit volume cost (cost per unit of volume) is lower as the tank volume increases so, for irrigation purposes, it is preferable to use one big tank rather than two or three smaller tanks. For domestic use, however, two tanks are recommended in the courtyard: one for storing higher quality roof water and another for the ground runoff. The dimensions for tanks of different volumes with circular sections are shown in Table 2.4.

Replication of new rainwater harvesting techniques in Gansu

The domestic RWH system described above, including the upgraded water cellar, was tested and demonstrated in a research and demonstration project conducted from 1988 to 1991. At the outset most of the farmers just wanted to see if the system really worked. Another major initial obstacle to promoting the new RWH system was caused by the fact that the GRIWAC team asked householders to build the tank inside their courtyards to prevent possible pollution by livestock and poultry; however, local people hold the superstition that if the earth is disturbed in the courtyard then the family may suffer misfortune. Consequently, only a few households were willing to participate in the first pilot project, despite a subsidy of $50 being provided. After one year of using water cellars located inside their courtyards no misfortune had been observed among the participating families. In fact they were named as 'scientific families' for their more open-minded acceptance of this new development. Rather than suffering bad luck, these families' water supply was much improved both in terms of quantity and quality. Consequently the following year there was a rush of households in the demonstration villages applying for subsidies to build pilot RWH systems in their courtyards.

The construction of water cellars was further promoted through a small-scale extension project after the findings of the GRIWAC research and demonstration project were accepted by the Gansu provincial government.

The provincial government distributed 4 million CNY ($500,000) annually to help farmers build RWH systems. At the end of 1994, the number of water cellars reached 22,280, with a total storage capacity of about 450,000 m^3. From mid-1995 to the end of 1996, with the support from the provincial government through the 1-2-1 Project, the number of water cellars jumped to around half a million and the beneficiaries amounted to 1.3 million. The 1-2-1 Project provided a subsidy to farming households to construct one tiled catchment and two water cellars, and supported micro-irrigation using rainwater on one piece of cropland. Since the start of the 21st century, other new campaigns have also promoted the further construction of water cellars. A 'Mother's tank' campaign initiated by the China Women's Federation and the China Women's Development Foundation helped Gansu build a further 8,620 *Shuijiao* and 2,475 tanks from 2001 to 2002, benefitting 63,000 rural people. Simultaneously, another campaign called 'Love given water cellar' donated a sum of 6.23 million CNY (about $750,000) to help 10,557 families of Zhuanglang and Wudou counties build 10,920 water cellars. This provided 28,400 people and 7,040 livestock with cheap and reliable water supplies from the systems. In addition, about 660 ha of cropland and 150 low-cost greenhouses took supplemental irrigation water from the cellars.

In the past 20 years, almost 2.9 million upgraded *Shuijiao* were built, of which a little more than half a million were for the domestic supply and the remainder for irrigation. The total number of beneficiaries from the domestic RWH systems is 1.4 million. The average storage volume for each cellar is about 30 m^3. For water cellars between 20–30 m^3, the cost is about $220–300. Government support provides 1.5–2 t of cement, equivalent to $90–120. The remaining 60 per cent of the input is contributed in kind by the householders in the form of labour and local materials. All the water cellars have been built by the householders themselves, with some guidance and support from water technicians from the local township and county.

Reasons for the rapid spread of rainwater harvesting systems

There are a number of reasons why water cellars have developed so rapidly in Gansu. The most important is that the water cellar brings real benefits to the farmer and the household. The upgraded systems ensure adequate water supply to meet total household water demand. In the past, local people only paid attention to the construction of the *Shuijiao* clay-lined water cellar but failed to consider the collection capacity of the catchment, which was generally insufficient to fill the tank. This discouraged the farmers from building tanks and gave them limited confidence in RWH in general. Once it was realized that RWH systems consist of the three indispensable components of collection area, storage reservoir, and water supply delivery system, numbers increased and improvements in performance quickly followed. In 1991, Gansu suffered a serious drought, yet a household in Xifeng after building its RWH system had enough water to meet its demand and even help neighbouring families through the difficult period before the rainy season.

The people of Gansu have suffered from lack of water for generations and water shortages have challenged the local authorities for decades. In the past, during droughts the government requested state enterprises and institutions to send water trucks to provide supplies to remote villages. Villagers had to wait in long queues for a single bucket of water that would have to last for several days as no one knew when the next truck would come! Since the introduction of the upgraded water cellars, women no longer need to worry whether there is water for cooking. Families have saved a great deal of time that was once used carrying water from distant rivers at the bottom of deep valleys, a burden that usually fell on the shoulders of women and children. A survey carried out by the Gansu Bureau of Water Resources shows that the provision of a rainwater catchment system saves households up to 70 days annually. This saving is critical as it means that many more children now have enough 'free' time to attend school.

Another major improvement resulting from the RWH programme has been in the quality of domestic water. In the past local people often had to drink dirty water contaminated with sediment from sources shared with domestic animals. In other cases they had to rely on saline groundwater sources with a high mineral content for their survival, despite its potentially harmful effects on their health. In many of Gansu's villages, some water sources used previously either lacked certain elements or had an excess of others resulting in water-related diseases, for instance, Kashin-Beck disease (Osteoarthrosis defomas endemica) caused by a lack of selenium and dental disease caused by the high fluoride content in the water. It is expected that the water quality improvements resulting from RWH will also have a positive long-term impact on these health problems.

The construction of the upgraded RWH systems has also led to other benefits. The hardened courtyard catchment areas can be used for drying straw and food. Householders can use surplus water from the domestic RWH system and the household's wastewater to irrigate their small vegetable gardens. This has allowed them to diversify their diet with a wider variety of vegetables that were previously impossible to grow due to the lack of water. Local sayings reflect the people's joy over changes the RWH systems have brought about: 'The dirty water becomes clean; the distant water comes close; and the bitter water becomes sweet' and 'When drinking water, it should not be forgotten where it is from'.

As the programme grew, the advantages of RWH became increasingly apparent to both the rural population and decision-makers alike. This process of increasing awareness was greatly assisted through the hosting of many forums, symposiums, and meetings attended by experts and officials, where the utilization of RWH on the Loess Plateau of Gansu was promoted.

Another important reason for the rapid development of RWH in Gansu was the enthusiastic acceptance it received from local people. A programme on this scale can only be completed within a few years with the full support of the majority of the population. Almost all the systems constructed over the past 20 years have been by the farmers themselves, with only limited or no instruction from local government technicians.

Since the RWH approach used in the programme is based on an ancient and indigenous technology it is well understood and its upgrading has been easily mastered by the local population. The technique is appropriate, practical, and easy to apply. Perhaps most significantly, because most of the inputs are provided by the householders themselves in the form of labour and local materials, the systems are affordable with the government subsidy.

The provision of cement represents the main government subsidy to the programme. The government, however, recognizes that full ownership of each domestic RWH system belongs to the household who built it. This sense of ownership ensures that householders have been keenly involved in the planning, design, and construction of the systems and are fully capable of operating and maintaining them sustainably.

The solid political support from the government has also been an essential factor in ensuring the rapid spread of RWH in Gansu. After seeing the success of the demonstration and extension projects between 1988 and 1994, the governor and provincial government realized the great potential of RWH for solving the domestic water shortages faced by the people inhabiting the Loess Plateau of Gansu. A further positive endorsement of the programme followed an investigation commissioned by the governor of Gansu Province into the use of RWH systems in Yuzhong County and a favourable article on these findings published in the Gansu daily newspaper. This, along with the impact of the 1995 drought, finally led the authorities to make the decision to initiate the 1-2-1 Project. The 1995 drought sped up the decision-making process and the government quickly decided to implement the 1-2-1 Rainwater Catchment Project. The first problem faced, however, was how to finance it. To achieve the goal of the 1-2-1 Project – reaching 1 million people and solving their drinking water problem – 109 million CNY ($13.4 million) of subsidies was needed. The government managed to fund almost half of this from its own sources but then had to launch a fundraising campaign with the media to source the balance. A total of 56.43 million CNY ($6.9 million) was raised from the public and various sources, including other provinces and municipalities as well as foreign sources, ensuring the project could proceed. To guarantee the funds were used as intended, careful monitoring both by the government and the community was undertaken, as well as a strict financial audit. Regulations relating to the use of project materials and funds along with rules for project staff were drawn up to ensure the efficient and open management of the project's implementation.

Relevant government agencies were mobilized to ensure the smooth running of the project. Guidelines and policies for the project were agreed. The initial priority was to assist villages that had no surface or groundwater supply and where rainwater was the only available water source. The number of eligible people in various villages, townships, and counties was determined, the subsidies for each village calculated, and the financial support provided. Staff from the relevant government agencies went to the countryside to promote, assist, and monitor the implementation of the project.

Despite the overall success of the programme a number of problems still exist. The total amount of water available to farmers is still relatively low. This does, however, vary and depends on the size of the household catchment area. In the 1990s, a typical household might have had a roof area of 50–60 m² and courtyard area of 100–150 m². In a normal year a standard 1-2-1 Project domestic RWH system would provide such a household with about 15–20 litres per person per day in a location with 400 mm mean annual rainfall. In a dry year (one in ten), the quota will drop to just 10 litres per person per day, and in those areas with mean annual rainfall of 300 mm or less, the daily quota will of course be even lower. In these circumstances only the most basic needs of providing drinking and cooking water for people can be met. The situation should improve in the future; if farmers extend their houses and pave or harden more of their courtyards, the amount of water supplied could be raised significantly by 50–100 per cent or more. A recent survey showed that already in some villages where the catchment areas are larger, households can run a washing machine two to three times a week using only rainwater runoff, and even take showers – something that they have never had before.

A further constraint is that most roofs have no gutters or downpipes, with rainwater just falling from the eaves into a concrete channel on the ground and then into the tank. The systems need to be standardized, yet this makes funding a problem as so far the investment of the government is only for cement. Most households buy solar cookers for boiling water (see Chapter 3), which makes it much safer to consume. Nevertheless, some people still drink the rainwater without boiling it, indicating a need for more health education.

CHAPTER 3
Water quality and the development of solar cooker technology

Water quality issues

RWH systems have dramatically changed the nature of rural water supply in Gansu Province, both in terms of the quantity and quality of water. Before the project, the main sources for domestic supply for households were traditional water cellars or pits where floodwaters would remain after heavy rain. In the traditional rainwater systems, most of the collection surfaces were straw-mud roofs, rural unsealed roads, or bare natural soil slopes. These would produce little runoff during light rain, while in heavy rain the runoff from these catchments produced a lot of silt and dirt, resulting in poor water quality. In other places, people often shared contaminated water in temporary ponds and pits with animals. In some areas, spring water provided a source but here high mineral content was a problem; for example, high fluorine concentrations that can cause skeletal fluorosis and other diseases.

The implementation of the new RWH systems using tiled roofs and concrete-lined courtyards for collecting rainwater for domestic use has helped to greatly improve water quality, with the rainwater stored in the new tanks being universally recognized as much cleaner and fresher than before. Although householders are now generally satisfied with the water quality provided by the RWH systems, in Gansu catchments are not regularly washed down by the rain owing to the dry climate. As a result the water collected from the catchments is still often technically defined as being of poor quality.

Water samples tested from water cellars show that both bacteriological indices (coliform bacteria and total bacteria) and turbidity frequently exceed the limits stipulated by the China National Standards. For new water cellars plastered with cement mortar or built with concrete the pH value can also be high, with some samples reaching pH 10, although this does normally reduce naturally within six months of construction. GRIWAC has suggested that farmers should adopt 'first-flush' systems, some of which have proved effective in bringing about water quality improvement in other countries. A first-flush system is a device to reduce the contaminants by draining away the initial runoff from the catchment in the first few minutes of rainfall. So far local people in Gansu have not accepted these systems. This is because since rainfall is infrequent and people do not know how long the rain will last or when the next rain will come they are very keen to collect what little they can whenever it rains. They also think that the catchment will still be polluted even if they do the first flush because the rain is so irregular that it might be one or two months until the next rainfall. The fact that they are unwilling to sacrifice even a small quantity of water by using a first-flush system indicates just how highly they prize every last drop of rain.

Several measures have been suggested to improve quality but for financial and other reasons most of them have not been put into practice. One of them has been to store the runoff from roofs and the ground in separate tanks. This practice has been adopted by some households when residents have been informed of the benefits, but for most households the cost of an additional tank, gutter, and downpipe is prohibitive. Householders do not like to add agents for sedimentation and disinfection, as suggested by the Gansu Institute for Water Conservancy, because of their smell and taste.

Nevertheless, between 2008 and 2010 GRIWAC worked on a project to develop a household water purification system using an approach involving a flocculation agent and ceramic filter. Trials of this system to date have been very promising. The cost of the system is still beyond the means of most rural families, so its widespread introduction would require some form of state subsidy. It is still undergoing field trials, the details of which can be found in Appendix 1.

Despite some concerns, since 1995 more than 1 million rural people have started to use rainwater as their main drinking source, and so far there have been no reported cases of diarrhoea or other waterborne epidemic diseases that could be attributed to harvested water. This could possibly be the result of the population building up natural resistance to relatively low levels of bacteriological contamination.

Boiling water from the tanks before consumption has also been widely practised by households, which of course helps to protect against any pathogens. This is mainly because the local people drink tea as their main beverage. However, fuel for boiling the water and for cooking is a problem. This area is famous for the so-called 'three shortages' (water, food, and fuel). In the past, farmers used to cut very scarce shrubs and grasses, and burned these with crop residues for cooking. This seriously depleted the soil of organic matter and fertility, and seriously damaged the whole ecosystem. To avoid this, Gansu provincial government started to popularize solar cookers in rural areas from the late 1970s onwards.

Technical aspects of the solar cooker

The principle of the solar cooker is to focus solar radiation onto a kettle by reflecting the radiation from a curved surface. Photos 3.1 and 3.2 show examples of solar cookers produced in Gansu.

Features of the solar cooker

The solar cooker consists of the cooker body, cooker framework, and direction-adjustment mechanism as well as the kettle (or other metal container) and a supporting frame.

The cooker body. The cooker body (or reflective dish) is the component for collecting, reflecting, and concentrating the sun's rays onto the base of the kettle. This has the shape of a parabolic curve. When parallel beams of sunlight strike it, the reflected rays are concentrated at the focal point of the curve, as shown in Figure 3.1. Further details regarding the method for drawing a parabolic curve can be found in Appendix 2.

The cooker dish can be made of many kinds of materials: cement mortar (using common cement or sulphate aluminium cement), magnesite, calcium carbonate, high density polyethylene (HDPE) composite, epoxy resin, or concrete. All these kinds of material are reinforced with glass fibre or sometimes with wire mesh/steel bars. A steel, aluminium, or cast iron parabolic curved sheet can also be used. In Gansu, cement mortar and magnesite are the most commonly used materials as these are low cost. Magnesite is a little more expensive than cement mortar, but it has higher strength so the thickness and thus the weight of the cooker can be reduced. However, magnesite has low resistance to water so it is not suitable for moist environments.

The cooker surface is covered with a reflective material – usually strips of small glass mirrors. The glass fragments are about 35 × 40 mm and are 2 mm thick and plated with aluminium on the back. These are stuck onto the curved surface of the cooker body using hot asphalt as adhesive.

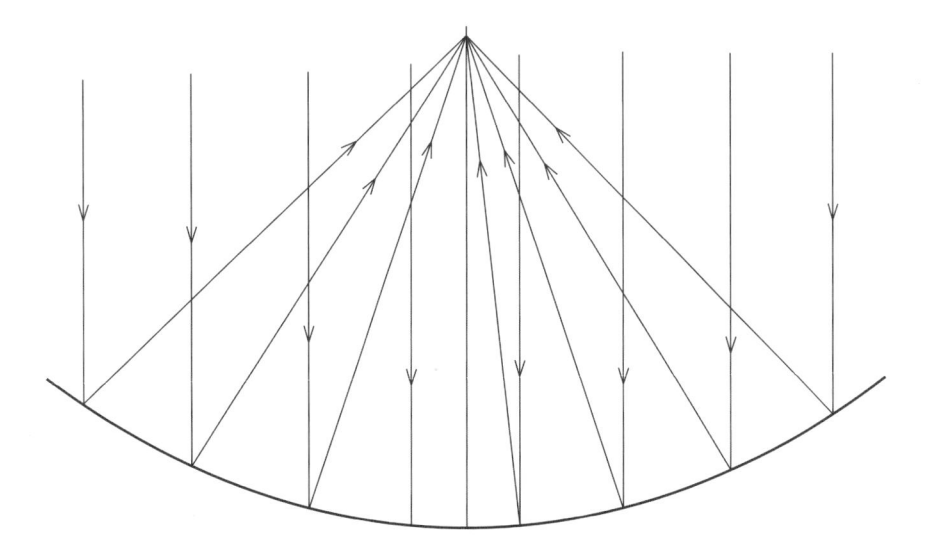

Figure 3.1 Reflection of parallel light beams from parabolic curve

Alternatively, aluminium can be used. This is a more costly option but is much lighter. The aluminium film is a polyester film coated with a very thin aluminium layer applied using a hot spray method. Its back is like adhesive tape so that it can be easily stuck on the curved surface. It is applied by cutting it into strips 6–7 cm wide and 20–30 cm long and then sticking these on to the reflecting surface. The aluminium film has a 10 per cent higher reflectivity than the glass mirror surface but has a shorter service life and is much more costly. Some aluminium cookers are made of several pieces that are easy to disassemble and convenient to package and transport and therefore suitable for mass production. While these are more expensive, it does make their rapid and widespread dissemination feasible, potentially on a global scale. A case study of the development of this type of solar cooker is presented at the end of this chapter.

The framework of cooker body (reflecting surface) and direction-adjustment mechanism. In order to work, the axis of the cooker must always be pointed towards the sun and there should be little or no cloud cover. As the sun moves across the sky from east to west, its bearing changes by about 15° per hour. Its elevation also changes through the day, rising from zero at dawn to a maximum at midday (this maximum value varies by about 47° between mid-winter and mid-summer).

A parabolic curve is most efficient for focusing the solar energy onto a kettle if its axis is kept pointing within about 20° of the sun, so the householder needs to adjust both the bearing and the elevation of the parabolic reflecting surface, to point it directly at the sun or a little ahead of the sun, before placing a kettle full of cold water at the heating focus.

The cooker body is quite heavy and has two roles: to support the cooker firmly and to follow the sun. To support this weight, a strong framework with a simple structure is necessary. This is designed so the cooker can be rotated about a vertical axis or on a wheeled stand to follow the sun from east to west. It can also be tilted up or down to track the sun's elevation. To provide this tilting mechanism the cooker body is installed on a horizontal shaft supported on the framework. There is a screw or rack structure linking the cooker body and the framework, to control the pitching movement.

The kettle and its supporting frame. The kettle (filled with water) is positioned to receive a concentrated beam of sunlight from the reflecting surface, which heats the water inside. The kettle is placed on a steel ring with a diameter of 30 cm positioned around 1.2 m above ground, i.e. suitable for the user to operate. The ring has to be located at the focal point of the reflecting surface and the plane of the ring should always be parallel to the ground whatever the tilt of the cooker body.

There are several kinds of frame to ensure these two criteria are met; two examples are shown in Photos 3.1 and 3.2.

Photo 3.1 Solar cooker with type I frame   **Photo 3.2** Solar cooker with type II frame

In the type I supporting frame, the steel ring supporting the kettle is supported on a vertical steel rod that is fixed on a base made of concrete or cast iron. On the reflecting surface is a slot allowing it to be tilted vertically. When the reflecting surface turns, the focal point will move but if the base of the kettle is large enough it will still receive focused sunlight but with slightly lower efficiency.

In type II, the supporting frame has an improved design. The steel ring, the front support rod, the bottom horizontal rod, and the distance between the centres of the steel ring and of the reflection surface each form four sides of a parallelogram (see Photo 3.2). When the reflection surface turns to follow the sun, the steel ring is always parallel to the bottom horizontal rod. Since the distance between the steel ring and the reflection surface is one of the sides of the parallelogram it remains unchanged and equal to the focal distance.

Factors affecting use of solar cookers

At latitudes higher than about 41°, the sun's rays are too weak and the solar cooker efficiency too low to make their use feasible throughout the year. So long as the sky is clear and the sun high enough in the sky, the cookers can boil water even when ambient air temperatures are well below freezing. Cloudy conditions and the hour or two after dawn and before dusk are clearly not conducive to effective solar cooker use. Gansu lies between latitude 32° and 43° north and annual sunshine hours reach over 3,000 in many parts, making

Box 3.1 Parameters of the solar cooker

For a proper design of the solar cooker, parameters to be determined include:

1. *The design solar elevation (angle).* The angle between the sun and the ground is called the solar elevation. This changes with latitude and the time. The maximum angle of the sun is highest in summer and lowest in the winter. In places with latitudes less than 23.5° North or South, the maximum solar elevation at noon is 90°. When the latitude is larger than 23.5°, then the maximum solar elevation equals 90°–(ϕ–23.5°) where ϕ is the latitude. When designing the solar cooker, the maximum solar elevation is taken as the highest elevation the sun attains at that latitude but not greater than 82°. The smallest angle is taken as 25° because if the solar elevation is below this the optical efficiency of the cooker will be too low for it to work.
2. *The operating height and distance.* The operating height and distance should conform to a human scale for convenience of operation when he/she cooks. The China national standard for solar cookers stipulates that the maximum height of the cooker and the maximum distance from the cooker surface to the kettle is not greater than 1.25 m and 0.8 m, respectively.
3. *The focal distance and the interception area.* The focal distance is the distance between the origin of the parabolic curved surface and the centre of the kettle. The interception area is the projected area of the curved reflection surface perpendicular to the sun's rays when the axis of the surface is parallel to them. The power of the solar cooker is proportional to the interception area and a large focal distance results in higher efficiency. However, too large an interception area will increase the cost and may make the operating height and distance between the kettle and cooker unpractical. On the other hand, if the focal distance is too short, the operating height and distance will be smaller, but the interception area could be insufficient to generate enough power. See below.
4. *The interception area.* The interception area is calculated with the following equation:

$$Ac = \frac{P}{I_d \eta_L} \qquad (3.1)$$

where P is the rated power of the cooker in watts, I_d is the direct solar radiation in watt/m², and η_L is the optical efficiency of the solar cooker. The rated power ranges between 800 and 1,500 W. I_d depends on the latitude – in Gansu it is about 800 W/m². The optical efficiency can be taken as 0.65. For example, in the Loess Plateau of Gansu, to get a solar cooker with power of 800 W, the interception area needs to be about 1.5 m²

---

conditions for solar technology generally very good, especially in the middle and eastern part of the province where most of the solar cookers are used.

Manufacture of the solar cooker

Production of the cooker body mould template. In order to make solar cookers in large numbers it is necessary to first make a mould on which the cooker body reflecting surface can be formed. For this, a template of wood or steel plate is required. To make the template, a half section of a parabolic curve should be drawn on the wooden or steel plate using the method described in Appendix 2 to generate the curve. The template should be wider than the designed reflecting

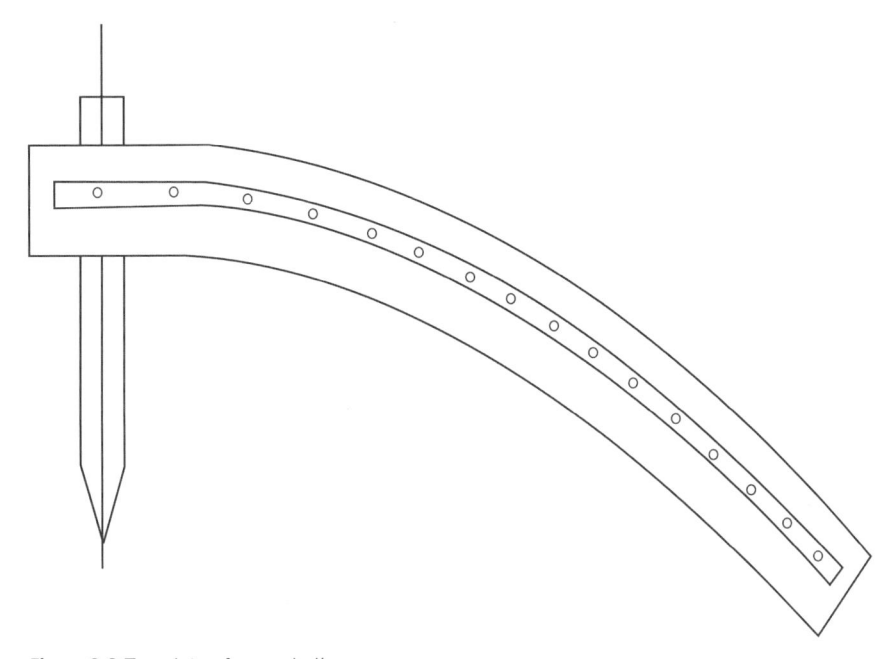

Figure 3.2 Template of a parabolic curve

surface by 20–30 cm. The section of plate should be carefully cut along the parabolic curve leaving a width of 20 cm of plate above the curve. The drawing and cutting must be as precise as possible to ensure the template represents an exact parabola. The template should be completed with a pivot around which the template can be rotated. Figure 3.2 shows a schematic drawing of the template.

Production of the mould. In order to reduce costs, the mould for manufacturing the cooker body (reflecting surface) can be built on the ground. The first step is to level and compact the ground. A half dome with a height of 30–40 cm is built first with compacted soil in the inner part and then with cast-in-place concrete on the surface with fibreglass for reinforcement. A pipe with an inner diameter a little larger than the pivot on the template should be anchored in the centre of the dome and the pivot of the template is inserted into the pipe with its lower edge higher than the dome top by a couple of centimetres. Then the top of the dome can be shaped to a parabolic surface by adding cement mortar on top of the dome while rotating the template. The excess mortar will be shaved off and extra mortar should be added where required to produce a surface that exactly conforms to the template. The surface of the mould then needs to be polished with sandpaper or other tools. Photo 3.3 is a picture of the concrete mould. The finished mould should be cured for several days by spraying with water and covering with plastic.

36 EVERY LAST DROP

Production of the cooker body (reflecting surface). The procedure for producing the cooker body is as follows:

1. When the concrete mould has hardened, a plastic film with a thickness of 0.01 mm is put on the mould to facilitate removal of the cooker body after it is produced. Alternatively a lubricant oil can be brushed on the mould.
2. An outline of the shape of the cooker body is drawn on the module and a wooden or steel surrounding frame (3 cm in height) with the same configuration placed on the concrete mould. Meanwhile, slots need to be made for the free steel framework that supports the cooker body and kettle (type I, Photo 3.1). A steel rod handle for lifting the cooker body should also be added at this stage.
3. Mortar with a thickness of 1 cm should then be plastered on the mould. The mortar mix should have a ratio of cement to sand of 1:2.5. A layer of wire mesh or fibreglass is placed into the mortar. The wire mesh or fibreglass should be smaller than the surround frame by 2 cm on all sides.
4. A second layer of mortar (1.5 cm thick) should then be plastered over the wire mesh or the fibreglass and pressed to form a close connection with the lower layer.
5. The main component of magnesite is magnesia (magnesium dioxide) mixed with a solution of magnesium chloride in the ratio of around 4:3 (by volume). Sawdust (6–8 per cent by volume), or very fine sand (15–18 per cent) can be added to reduce the weight. Since the magnesite is alkali, wire mesh cannot be used as reinforcement as this will corrode. Instead, anti-alkali fibreglass is used. The total thickness of the magnesite is 1.2–1.5 cm. A layer of magnesite 1 cm thick is firstly plastered on the mould. Then the fibreglass and the upper layer of magnesite are added. Photo 3.4 shows the process of how the layer of magnesite is applied on the mould.
6. After the cooker body has hardened, it can be removed from the mould. For cooker bodies made of cement mortar this can be done after 2–3 days, depending on the temperature. These should then be kept in a

Photo 3.3 Mould with cast of the reflection surface

Photo 3.4 Plastering the first layer of mortar

Photo 3.5 Sticking on the mirrors   **Photo 3.6** The completed solar cooker bodies

moist environment for at least 10 days by sprinkling with water and/or covering with plastic or damp sand. Cooker bodies made of magnesite can be removed after 24 hours and placed in a cool and ventilated environment.
7. The cooker body is then placed on a temporary frame and asphalt is brushed on as adhesive for the small pieces of mirror (Photo 3.5) to complete the cooker body (Photo 3.6).

The cooker body can also be made of cast iron, steel, or aluminium plate in the factory, although the process is not described here. An example of such a product is shown in Photo 3.7.

Proper use of the solar cooker

The proper use of the solar cooker is important as it influences its efficiency and service life, as well as for safety reasons. To ensure proper use:

- Operation of the cooker should only be by adults. Young children especially should be kept away from the cooker.
- The person using the cooker should stand behind it, i.e. behind the high edge.
- To minimize the risk of fire and other harm care should be taken avoid the concentrated reflected light beam focusing on windows, houses, trees, animals, washing etc.
- When the cooker is not in use it should be turned to a vertical position.
- Clothes and other materials should not be placed on the kettle frame.

To maintain the cooker:

- Snow, dust etc. should be removed from the reflecting surface.
- Glass mirrors can be wiped off with or without using water, but water should not be used for cleaning an aluminium-coated cooker.
- The cooker should be covered during rain and cookers made of magnesite should always be kept dry.

Popularization of the solar cooker in rural Gansu

The first solar cooker was developed in 1956 in Shanghai. Research and development of the solar cooker in Gansu started in the 1970s and the demonstration and extension first took place in Dingxi and Linxia districts. By 2010, there were 10 solar cooker manufacturers in Gansu Province, each capable of producing more than 10,000 sets per year. The total provincial production capacity amounts to 300,000 sets annually.

The solar cooker can boil 3 litres of water within 20–30 minutes in sunny conditions. When preparing meals, people first use the cooker to boil water and then pour the water into a pot for cooking. A solar cooker produced locally costs about 200 CNY ($30) and is affordable to most rural families. The service life can be as long as 8 to 10 years if used properly.

The introduction of solar cookers has produced significant economic, social, and environmental benefits in rural Gansu. Investigations into the impact of the cookers have revealed that each solar cooker can save households the equivalent of 750–1,000 kg of firewood or 550–800 kg of coal each year. Taking the cost of the firewood and coal at 0.3 CNY and 0.5 CNY per kilogram, respectively, the investment in a solar cooker can be returned within one year. Use of solar cookers also saves the time and labour of cutting firewood and crop residues and energy required for its transportation – tasks mainly undertaken by the women and children. Other benefits of the solar cooker include the purification of drinking water through boiling it, thereby reducing exposure to waterborne diseases. The cookers also provide warm water for livestock in wintertime. In addition, the rural-based solar cooker factories provide scarce job opportunities, often for hundreds of people.

The reduced removal of crop residues such as straw for fuel and the reduction in the cutting of trees have several direct benefits to the environment. The increased amount of organic material in the soil improves soil fertility and helps prevent soil erosion. The reduction in the amount of coal burnt, still a very common fuel in rural China, also has implications for reducing greenhouse gas emissions. It is estimated that the 578,000 solar cookers installed in Gansu by the end of 2006 annually produce energy equivalent to 170,000 tonnes of standard coal or carbon dioxide emissions of 460,000 tonnes.

Case study: Qiang Li and his Huaneng Solar Energy Co. Ltd

Local entrepreneurs have played a vital role in the rapid development and spread of solar cooker technology in rural Gansu and the story of Mr Qiang Li illustrates this well. After Li graduated from the Department of Mining Engineering at Ningxia University he went back to his native Gansu and worked in the state-owned Lanzhou Mining Machinery Factory. After some time he thought he would establish his own business, so he resigned his job and set up to be a building contractor. In 1992 he met with Mr Shimin Li, Deputy Director of the Lanzhou Research Institute for Solar Energy and started to gain some knowledge about solar energy. He could immediately see the potential of this technology, especially in rural Gansu due to the lack of fuel in much of the province but its abundance of solar energy. With technical support and technical design information for the solar cooker provided by the Institute, Qiang Li invested all his money into setting up a small-scale solar cooker workshop with yearly production capacity of just 2,000 sets.

The first challenge Li faced was when he started to promote the product in villages. To demonstrate the effectiveness of the cookers he conducted a demonstration in front of members of the community by putting a piece of cardboard at the focal point of the cookers. Within a few minutes, the cardboard started to burn. The villagers, however, still did not believe that the cookers would work, believing this was just some sort of magic trick. So Quiang Li lent solar cookers to the families of 10 village leaders for one month to use free of charge. After a month, he returned to collect the cookers, but the families would not part with the cookers and instead insisted on buying them. They all agreed the solar cookers were very useful and well worth the price. From that point on the business started to develop, while at the same time, thanks to the promotion activity of the Solar Energy Institute, the Gansu government also started to recognize the benefits provided by the solar cookers in terms of both improving rural life and conserving the environment. In the mid-1990s the provincial government started a subsidy programme to assist rural families when buying solar cookers.

Photo 3.7 Foldable solar cooker

In 1997, families in rural Gansu installed 60,000 solar cookers with the help of this initiative. The provincial government still purchases around 80 per cent of the solar cookers produced at Qiang Li's factories so he never worries about the market for his products. Currently, the company has an annual production capacity of 140,000 units. In addition to the solar cookers, the company also produces solar water heaters, solar ovens, and other energy-saving products. He has become a famous entrepreneur in Dingxi County in Gansu.

Over the years Li has made many improvements to the solar cooker technology. His first solar cooker was made of cement mortar. Li found this too heavy and another problem was that the cement corroded the mirrors over time, reducing their reflectivity significantly after several years. After consulting experts, he developed designs using other materials such as magnesite and cast steel, as well as steel and aluminium plate. Using these materials the weight of the cooker can be reduced by more than half. Li also developed a special type of reflective mirror with a thickness of just 1.8 mm, as well as using aluminium-plated film. For ease of delivery, he developed a foldable aluminium solar cooker weighing just 15 kg, which can be packaged in a small box. Due to his attention to detail, ensuring the high quality of his products, he has gained the confidence of both his customers and the government. Under the new national policy of developing a green economy, he will enlarge his business and contribute to the green energy economy.

PART II
Rainwater harvesting and sustainable agriculture

CHAPTER 4
Development and replication of low-cost greenhouse designs

Background

Initially the focus of the RWH project in Gansu was on improving rural domestic water supply. The rapid adoption of the upgraded water storage tanks described in Chapter 2 was widely replicated through the 1-2-1 programme and has helped to solve the severe problem of periodic water shortages that had plagued the region for generations. The full impact of the RWH project has turned out to be far greater than just the provision of a reliable and accessible domestic rural water supply. In the early 1990s, GRIWAC started to experiment with the use of relatively small quantities of harvested rainwater for irrigation purposes. These initial trials involved supplementary irrigation to rain-fed crops at a number of critical periods in the growing season when the plants would otherwise experience severe water stress. This method was then piloted with local farmers and the impacts were impressive. The approach involved using rainwater stored in the same types of subsurface water cellars (*Shuijiao*) that were used for domestic supplies, and applying small amounts of irrigation water to each individual plant at key periods in the growing cycle when supplementary irrigation was most needed.

This approach not only raised the crop yields significantly but also allowed farmers to greatly diversify the range and types of food and cash crops they could grow. Before the introduction of RWH for supplementary irrigation, 95 per cent of the cultivated land area grew cereals. Due to the low market value of cereals and the fact that most cash crops were impossible to grow, most families remained poor well into the 1990s. Most of the Loess Plateau is above 2,000 m altitude and is subject to severe frosts in winter as well as low rainfall. So prior to the RWH project the majority of people could only grow cabbages and potatoes and seldom ate what they called 'fine' vegetables. These comprise tomatoes, peppers, egg plants, herbs, and carrots, which since the introduction of rain-fed micro-irrigation with RWH have become possible to grow throughout Gansu, even in remote mountainous areas.

The introduction of low-cost greenhouses has played an important role in this development as these have helped to provide a moderate micro-climate in which a much wider range of crops can thrive. The development of demonstration projects by GRIWAC, involving the construction of 16 simplified greenhouses using plastic sheeting, began in the Beishan mountains of Yuzhong County, one of the most impoverished regions of Gansu as recently as the early 1990s.

Photo 4.1 Plastic shed – the first greenhouse in the agricultural history of Beishan mountains in Yuzhong in 1995

Photo 4.1 shows one of the first greenhouses constructed in this project. This simple structure consists of a bamboo frame covered with plastic sheeting. The success of the technology has been dramatic, with over 100,000 greenhouses being constructed since this modest beginning. Although the technology may be simple, the impact it has had on the lives of farming families and the community as a whole has been impressive.

This change has not been straightforward, since due to the remoteness of many rural communities in the mountainous areas of Gansu it has not always been possible to take surplus produce to markets. The owner of the greenhouse shown in Photo 4.1, however, shared her surplus with relatives and received some of their surplus grain in exchange. Since the 1990s there has been an improvement in rural roads and as the economy has grown more marketing opportunities have sprung up, resulting in steadily higher economic benefits for rural greenhouse producers.

Studies conducted by GRIWAC have shown that the cost of constructing a simple greenhouse with an area of 350 m^2 is about 10,000 CNY ($1,450), not including the labour provided by the farmer. In Gansu, at least two harvests can be expected each year, yielding around 4,500 kg of vegetables with a market value of around 7,200–9,000 CNY ($1,040–1,300) annually, providing a return on the investment within 18 months. These simple greenhouses, which generate a significant proportion of their micro-irrigation water needs using rainwater runoff, have been an important contributor to income generation and poverty alleviation across large areas of the Loess Plateau in Gansu Province.

Box 4.1 Greenhouse development in Daping village

Daping village in Qinglan township, Dingxi County, is one of the poorest and driest villages in the county. Until the 1980s the only water source in the valley was a turbid and bitter-tasting spring producing inadequate water to meet the demand.

Villagers had to get up early and queue for long periods when fetching water. During the 1990s the spring's flow decreased, disappearing altogether during the severe drought of 1995. In response the villagers built RWH systems for their domestic water supply.

In 2003, Daping was chosen as one of the pilot communities for the 'New Village Construction Project' in Dingxi County. With the support of the local government, the farmers updated their houses and courtyards. The project allowed for the construction of a further 369 water cellars, each with capacity of round 50 m^3. An asphalt road was also built from the highway to the village and this became the main rainwater catchment for the village's RWH irrigation schemes. The state-owned Jinchuan Nickel Refinery Corporation funded a plan to construct an additional 41 greenhouses in Daping, each with an area of about 600 m^2, as well as a further 30 water cellars with a mean capacity of 64 m^3 for supplementary irrigation. Drip and pipe irrigation systems were installed in all the greenhouses. The stored rainwater allowed for between five and eight water applications before each harvest, using 10–15 m^3 of water for each application. Table 4.1 shows the irrigation water use under different irrigation methods for different crops.

In 2007, a study was undertaken to monitor the impact of the project. Based on a sample of six greenhouses it was found that the total production was 32,900 kg, with a market value of 56,050 CNY ($8,100). The total irrigation water volume used was 1,258 m^3, resulting in crop productivity per unit of water of 26.1 kg/m^3 or 44.6 CNY/m^3 ($6.45/m^3). When compared to the rain-fed cultivation of wheat, with a value of only 3,800 CNY/ha ($550/ha), the value of greenhouse production was 35 times higher.

Table 4.1 Irrigation water use for different irrigation methods and crops

Crop name	Irrigation area (ha)	Irrigation method	Number of applications	Irrigation water amount (m^3)	Irrigation amount for unit area (m^3/ha)
Tomato	0.76	Drip	5–6	1,185	1,559
Cucumber	0.24	Drip	5–8	386	1,608
Kale	0.24	Furrow	5–6	401	1,671

Research and field tests have shown that greenhouses using high efficiency micro-irrigation drip systems can generate consistently high profits. The value generated for every cubic metre of rainwater used was around 2–5 CNY for field crops, while in greenhouses planted with vegetables it is about 30–50 CNY. In some villages, greenhouses have been used for fruit trees and here the benefits were even higher than those from vegetables.

In the Liuping township of Qin'an County, a villager planted nectarine trees in his greenhouse. This allowed earlier maturity and ripening of the fruit when compared with fruit grown in the conventional way. The farmer was thus able to harvest the fruit two to three months earlier than most other fruit growers when the price was between 6 and 20 times higher than that around the usual harvest time. In this case, the production value per cubic metre of

Table 4.2 Yields and values of different crops from the greenhouse

Crop	Supplemental irrigation water (m³/ha)	Yield (t/ha)	Yield/ irrigation water (kg/m³)	Price (CNY/kg)	Value produced (10³ CNY/ha)	Value/ irrigation water (CNY/m³)
Cucumber	3,600	110	30.5	3	330	91.5
Tomato	1,950–3,750	92–122	43–48	1.8	166–220	64.5–86.4
Melon	750	30	40	6	180	240
Sweet melon	900	30	33.3	6	180	200

Note: 8.3 CNY = US$1

Table 4.3 Yield versus quantity of irrigation for cucumber crop

Irrigation quantity (mm)	Yield (kg/ha)	Yield/irrigation quantity (kg/m³)
100	79,267	79.3
200	100,088	50.0
300	127,586	42.5

Source: Data courtesy of Prof. Guo Xiaodong, Vegetable Institute, Gansu Academy of Agricultural Sciences

rainwater amounted to 150 CNY. Table 4.2 shows the yields and values of different crops from the greenhouse.

The results of trials on the RWH irrigation systems in the greenhouses have shown that the highest irrigation efficiency is obtained when the smallest amount of irrigation water can be used over the greatest area. The test results in Table 4.3 show the yields for a crop of cucumbers subject to different amounts of irrigation.

We can see from Table 4.3 that the yield per unit of irrigation is highest when less irrigation water is applied. From the perspective of maximizing the water use efficiency (WUE) of rainwater, a strategy involving the use of lower quantities of irrigation water should be adopted.

General introduction to the greenhouses in Gansu

There are two types of greenhouse in Gansu: the arch-type plastic shed and the solar-heated walled greenhouse (see Figures 4.1 and 4.2). The arch-shape greenhouse with the plastic sheet roof works like a conventional greenhouse, raising the inside temperature on sunny days by preventing heat loss in the form of air convection. At night it retains this warm air and protects crops from frosts. It can only work in the parts of Gansu with the least severe winter temperatures or only in spring to autumn in the colder regions of the province.

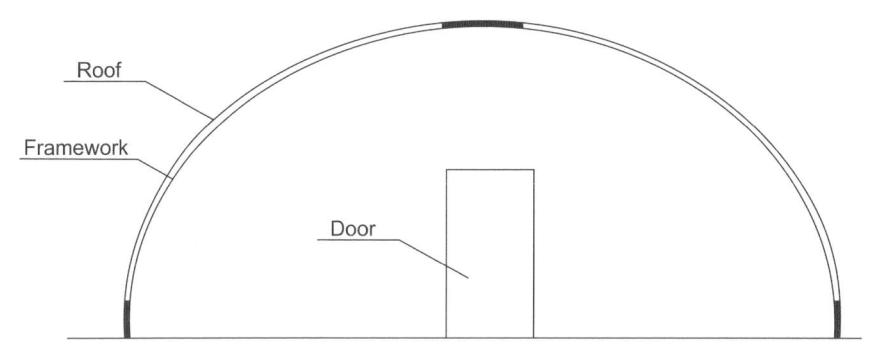

Figure 4.1 Arch-type plastic greenhouse

Figure 4.2 Solar-heated walled greenhouse
Source: Courtesy of Prof. Guo Xiaodong, Vegetable Institute, Gansu Academy of Agricultural Sciences

The solar-heated walled greenhouse not only receives solar radiation that raises the air temperature but, due to the high heat capacity of the thick rear wall, it also dissipates the heat absorbed in the daytime to help maintain an even temperature at night. The back wall also serves as excellent insulation. This greenhouse design works well in the wintertime in cold and even very cold regions. The walled greenhouse is able to retain more heat and maintain more even temperatures overnight. It thus produces two to three harvests a year, giving greater benefits to the farmer than the simple but much cheaper arch-type design. In Gansu the structural design of the walled greenhouse

48 EVERY LAST DROP

has been optimized to reduce its cost while at the same time improving its performance.

To simplify matters, we hereafter use the term 'plastic shed' for the arch-type greenhouse and simply 'greenhouse' for the walled-type shown in Figures 4.1 and 4.2, respectively. The plastic shed is composed of an arched steel or bamboo framework covered by a 0.15–0.2 mm thick plastic film. Vertical panels at each end, also made of plastic film, act as doors. In the walled greenhouse the roof is divided, with the large front curved plastic section for lighting, and a back roof, above the rear wall, providing insulation. A ditch about 50 cm deep is dug along the front of the greenhouse to reduce cold penetration from the frozen ground in winter. The following sections deal mainly with the more commonly used walled greenhouse (Figure 4.3).

The functions of the greenhouse in improving the crop-growing conditions in Gansu are primarily due to the following.

The protection of crops from frost damage

Temperatures on the Loess Plateau of Gansu Province in wintertime can be very low, often falling to -20°C or below. The walled greenhouse can provide effective frost protection even at these low temperatures. Despite the number of days with frost varying from 140 to 240 in different parts of Gansu Province, the climate is otherwise favourable for greenhouse production due to the high sunshine hours. Most years, sunshine hours in Gansu range between 3,000 and 3,500. In winter, sunny conditions prevail for 50–60 per cent of the time during daylight hours. Winter sunlight irradiating the greenhouse ranges from 20,000 to 40,000 lux over six hours on a sunny day. Under these conditions, the maximum internal air temperature reaches over 25°C for three to five hours. The thick back wall absorbs the heat from the sun, raising the temperature in the day and releasing this heat at night, keeping the internal

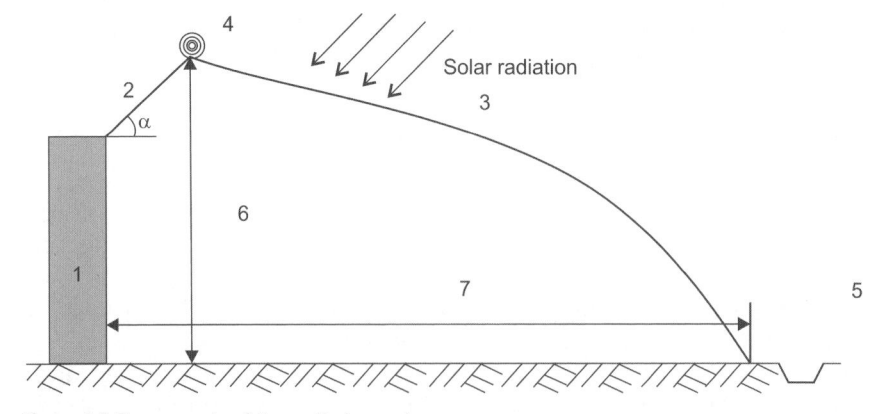

Figure 4.3 Components of the walled greenhouse
Note: 1 is back wall; 2 rear roof; 3 front roof; 4 heat insulation cover; 5 ditch to prevent lateral frost penetration; 6 height; and 7 span.

temperature relatively warm. The high specific heat capacity of the thick mud and brick walls makes them ideal for temporary storage of solar energy.

Additional measures are also used to reduce heat loss at night in the cold season, such as covering the roof with straw mats that can be rolled down over the plastic roof manually or using a motorized roller. This allows the internal night time temperature to be maintained at no less than 10°C. Only in extremely cold weather may it be necessary to heat the greenhouse for short periods to protect against the cold. Vegetable crops can therefore be grown normally even in the wintertime. In the summer, the temperature inside the walled greenhouses can sometimes get too high. A commonly used measure is to roll up the plastic film on the front roof to improve ventilation. Alternatively, a sprinkler system can be installed in the greenhouse to reduce the temperature by spraying crops with a limited amount of water.

Harvesting rainwater for greenhouse irrigation

Since the roof of the greenhouse is made of plastic sheeting, the rainwater runoff coefficient is typically in the range 0.85–0.9 (85–90 per cent). Consequently, the greenhouses are able to collect much of the irrigation water they need from their own plastic-film roofs. The results of field tests from four greenhouses are shown in Table 4.4.

From Table 4.4 it can be seen that, despite the low rainfall, the water collected from the greenhouse roof comprises about 40 per cent of the irrigation water required for two harvests a year. The additional water needs can be met by harvesting rainwater runoff from purpose-built cement catchment aprons adjacent to the greenhouse or from nearby roads. This water is stored in a large underground water cellar where it is protected from freezing in winter.

In situations where there are several greenhouses arranged in rows, shading problems can be avoided by leaving a space between each row. Gansu's relatively high latitude (35–37°N) results in the winter sun being quite low in the sky even at midday, so the distance between greenhouses should be around 8 m. In the higher latitude northern part of the province it should be

Table 4.4 Rainwater collected from the greenhouse roof and annual water demand

Owner's name	Greenhouse roof area (m^2)	Water collected from roof (m^3)	Additional water collected (m^3)	Total collected water (m^3)	Percentage of roof water
Wang Jianlin	285	45.1	69.9	115	39.2
Wang Weifeng	304	50.9	63.1	114	44.4
Wang Dongfeng	285	49.1	95.9	145	38.2
Ma Ziran	247	39.6	50.4	90	42.2
Mean	280.3	46.2	69.8	116	42.3

Source: Data courtesy of Prof. Guo Xiaodong, Vegetable Institute, Gansu Academy of Agricultural Sciences

Photo 4.2 Group of greenhouses in Dingxi County

slightly more. This strip of land is normally lined with concrete to provide an additional catchment area from which rain can be harvested. Where there are rows of greenhouses, the space between them is usually used as a road that is paved with asphalt and can also be used as an additional catchment. Photo 4.2 shows a group of greenhouses in Dingxi County. The space between the greenhouses both helps avoiding shading and provides some additional rainwater catchment area.

In parts of Gansu with annual precipitation of 550 mm or more, the rainwater collected from the greenhouse roof can provide enough water for the greenhouse to be self-sufficient for irrigating most types of vegetable. In areas where the annual rainfall is lower than 550 mm, additional rainwater collection areas will generally be needed to supplement the irrigation water requirements for vegetable crops.

Regulation of the humidity in the greenhouse

In situations where the humidity inside the greenhouse becomes too high for the plant growth, the plastic sheet roof can be rolled up to improve ventilation.

Design features of the greenhouse
Parameters for greenhouse design

The span. The span is the distance from the inner side of the rear wall to the toe of the front roof (Figure 4.3). The size of the span has a significant

Table 4.5 Span of greenhouse related to the outdoor lowest temperature

Outdoor lowest temperature (°C)	< -20	-20– -15	-15– -8	> -8
Span (m)	7	7–7.5	7.5–8	8–9

influence on the amount of sunlight received, the internal temperature, and the growth rate of the crop. If the height of the roof is fixed then a large span will reduce the incidence angle of sunlight and thus the internal temperature. A large span is also disadvantageous for maintaining the internal temperature after sunset due to the larger surface area available for heat loss. However, if the span is too small it decreases the land use efficiency (ratio between the crop area and the total occupied land) and makes the horticultural operations less convenient. To meet the required crop demand for sunshine and to ensure sufficient solar heating to maintain plant growth, prior consideration for an optimum design is required. The span of greenhouses in Gansu is related to the minimum average outdoor temperature as shown in Table 4.5.

Height. Height is the vertical distance from the roof ridge to the ground. For a fixed span, a low height results in a reduction of the incidence angle and the area of lighting as well as the space in the greenhouse. However, if the greenhouse is too high then the building cost will be increased. Furthermore, the greater the height, the larger the plastic roof area, leading to greater heat loss at night, especially in winter. An appropriate ratio between the span to the height has been established to avoid this problem. For cold regions like Gansu, the recommended span versus height measurements are shown in Table 4.6.

Length. The length is the distance between the two gable walls located at the east and west ends. While a greater length reduces the effect of sunshade by the gable walls, if the greenhouse is too long this will reduce ventilation and inconvenience operation. A length of around 60 m is generally found to be a suitable compromise.

Orientation of greenhouse. In almost all cases greenhouses in Gansu face south in order to receive as much solar radiation as possible. In the cooler regions of the province, mornings are much colder than evenings. As a result, in winter, the straw-mat insulation on the greenhouse roofs is not removed in the morning. Also, in order to fully maximize the benefit of the afternoon sunshine, greenhouse orientations in these situations are angled 5 to 10° west of south.

Table 4.6 Height versus span of the greenhouse

Span, m	7	7.5	8	9
Height, m	3.2–3.5	3.3–3.7	3.5–3.8	3.8–4.2

The angle and shape of the front roof. The transparent plastic roof is designed to allow solar radiation into the greenhouse. The inclination of the roof varies and is measured as the angle between the roof and the horizontal. To receive the maximum solar radiation, the roof should be as perpendicular as possible to the sun's rays. When this is so there is virtually no reflection and almost all the sun's radiation will pass through into the greenhouse. However, compared to the tropics, the sun, even at midday, can be relatively low in the sky in northern China, especially in the winter. At this time, the front roof would have to be close to vertical if the sun's rays were to be perpendicular to the roof surface and this would require an unrealistically high rear wall.

Since, the sun's angle changes throughout the day as it moves across the sky, the rays cannot always be nearly perpendicular to the roof. A computation model was developed that can simulate the solar radiation received for different solar angles at different times of the day. This simulation model suggests that the best design for producing the optimum internal climate has a shape between a parabolic and circular curve. Alternatively, it can be in three straight sections with angles from the horizontal of 15° for the upper, 30° for the middle, and 70° for the lower part, respectively.

When determining the slope of the greenhouse front roof, other factors in addition to light efficiency need to be taken into consideration. Easy drainage of rainwater, for example, is important as no water should remain in any depressions on the roof. The plastic film of the roof is easily held in place by ropes preventing it from being blown by the wind. An appropriate space at the front toe of the greenhouse is needed to facilitate cultivation activities as the roof becomes low on this side.

Thickness and height of the back wall. Thickness of the back wall is determined by the local climatic conditions and the wall materials. The thickness can be calculated by the thermal conduction equation and the thermal conduction coefficient of the wall materials to first determine the required thermal resistance of the back wall and thereby the wall thickness. If the back wall is made of soil, then the thickness can be determined by taking the mean depth to which the soil is frozen and adding 50 cm. If the wall is composed of brick or insulation materials such as dry soil, slag, pearlite, or polypropylene foam in between the brick layers, the thickness can be 40–80 cm less than for an earth wall. For example, the earth wall is usually 130–140 cm thick in Lanzhou, while a composite wall around 60 cm composed of polypropylene foam (8–12 cm thick) between two layers of brick, each 24 cm thick, would suffice. The height of the back wall is normally between 2 and 2.6 m.

Inclination of the rear roof and its width. The inclination and width of the rear roof has an important effect on the heat conservation performance of the greenhouse. Since the rear roof is not transparent if its inclination to the horizontal is too small then it will partially shade the back wall (see Figure 4.3 and Photo 4.3). The inclination of the rear roof should thus be a little larger than the maximum sun elevation angle at the winter solstice (in the northern hemisphere, this is usually on 21 or 22 December). In Gansu, the inclination of the rear roof is taken as 35–45°. A wider rear roof with good insulation will increase the heat preservation of the greenhouse, however, it will also increase the cost. In Gansu the width of the rear roof usually ranges from 1.2 to 1.6 m.

Structure and materials of the greenhouse

The supporting framework. The roof of the greenhouse is supported by the ground at the front. At the back support is provided by multiple columns plus the back wall, a single column plus the back wall, or the back wall only. The greenhouse design with multiple columns has the advantage of a lower construction cost and was widely used in the early stage of greenhouse development. However, the columns block sunlight and occupy space that could be used for cultivation so this design is no longer used. Nowadays, the common type of supporting framework is either with a single row of columns or no column at all. Figure 4.2 is an illustration of the greenhouse with a single row of columns. Photo 4.3 shows the greenhouse without columns.

The columns are mostly made of prefabricated reinforced concrete. Steel pipes with large diameters are used only occasionally due to their high cost. The reinforced concrete columns are 12×12 cm with four longitudinal steel rods of 8 mm diameter and 4 mm steel wire used for the stirrups at 15 cm intervals. The grade of concrete is C20 or C30.

Photo 4.3 Greenhouse without columns

The roof framework consists of two different truss types. A steel truss is most common. When a single row of columns is used, an arched steel truss is used for the front roof and a simplified structure can be used for the rear roof. When there are no columns, the steel truss should be used for both the front and rear roofs. The shape of the truss for the front roof is a combination of circular and parabolic curve and the shape for the rear roof is linear. The steel truss is made of steel pipe and/or steel rods. The interval between trusses is about 1 m. Three or four steel pipes/rods are installed in the longitudinal direction to stiffen the trusses at their upper joints.

To reduce the cost, a combination of steel trusses and bamboo rods are used. The trusses are erected every 1.8 m, while 4 mm steel wires are strung across the trusses every 45 cm, extending beyond the east and west gable walls and then anchored into the ground. Then two bamboo struts 15 mm in diameter are put between each pair of trusses. For the rear roof, a large diameter bamboo rod is put along the greenhouse ridge and timber rods are supported from the bamboo to the back wall every 60 cm.

The rear roof is composed of multiple layers, starting from the bottom with a waterproof plastic layer, then a structure for load bearing, a layer for heat insulation, and a second waterproof layer. In the past straw and dry soil were used for heat insulation but in recent years these have been replaced by polystyrene foam.

The transparent roof materials. The material used for the front roof is plastic film, including polyvinyl chloride (PVC), polyethylene (PE), and ethylene vinyl acetate (EVA). In recent years, several plastic films with special properties have been developed to improve performance:

- EVA plastic film. This is a long-lasting membrane with the property of preventing drips from forming inside the greenhouse. It is manufactured by adding a plasticizer, an agent for ageing resistance, and an agent for preventing drip formation. This is achieved using an organic coating material on the surface of the film, which also helps to dust proof it.
- Multi-function PE composite plastic films. These are made of three layers with a total thickness of 0.08–0.12 mm. During the manufacturing process, an ageing resistance agent, an insulation agent, and plasticizer are added in the upper, middle, and lower layers, respectively. The first gives the outside surface of the greenhouse better antioxidation properties. Heat insulation performance can be improved with the addition of the agent in the middle layer, while the plasticizer on the inside surface prevents water drips.
- Another kind of composite plastic film is made using LDPE (low-density PE) or EVA with the addition of anti-oxidation and dust prevention agents on the surface, a high vinyl acetate content EVA with heat insulation agent in the middle, and low vinyl acetate content EVA on

the inner surface with insulation and drip prevention agents added. This composite has the advantages of higher transparency, higher resistance to cold temperatures and mechanical impact, better drip-prevention properties, and less susceptibility to infrared radiation. This increases its strength and makes splits rare. It also has higher transparency, better heat insulation properties, and a longer service life.
- Multi-function thin film. This is made of PE resin with added multiple agents. The thickness of this film is only 0.05 mm. The scattering of light inside the greenhouse can reach 54 per cent of the total incident radiation so the light is distributed on the crop very uniformly, which benefits crop growth. The mechanical strength of the film is higher and the service life is longer. This kind of film also improves the heat conservation of the greenhouse and raises the internal air temperature by 1–4.5°C.

Heat insulation cover materials for the front roof. The front roof is the surface through which sunshine enters the greenhouse in the daytime but also where the major heat loss takes place at night. It is therefore necessary to put covering materials on the roof during the night to reduce heat loss during the colder months.

Traditional cover materials include straw mats, paper coverings, and cotton quilts. The straw mats are woven using wheat or rice straw and have a low thermal conductivity that can reduce heat loss by up to 60 per cent. The paper covering uses four to six layers of strong brown craft paper that are put under the straw mats. The role of the paper is to stop heat penetration through the voids between the mats so the inside temperature can be raised by 3–5°C. The straw mat covering can be seen in Photo 4.2. In the picture the cover is rolled up.

The shortcomings of these traditional cover materials include their heavy weight, susceptibility to degradation and rot, and the difficulty of ensuring their high quality of manufacture. The updated materials developed in recent years comprise an exterior layer made from waterproof textile, plastic film, plastic non-woven textile, or aluminium-coated film, and an inner core made of a polymer fabric, needle-punctured fabric, waste wool, or a polymer foam. The thermal conductivity of these new cover materials is similar to the traditional materials but, being more airtight, their heat insulation performance is better. In terms of heat conservation performance, these new cover materials are still not ideal. The goal is to develop a material that can limit heat loss to less than 20 per cent at night. Photo 4.4 shows the cover made of polymer materials.

The covers are rolled up during the day and unrolled over the front roof at night. The operation is usually done manually, but in recent years a machine has been developed for this operation. The cover material can then be rolled up or unrolled simply by pushing a control button and it can even be computer operated. Photo 4.5 illustrates this automated system powered by a small electric motor.

Photo 4.4 Cover made of plastic tarpaulin
Source: Courtesy of Prof. Guo Xiaodong

Photo 4.5 Device for furling covering material
Source: Courtesy of Prof. Guo Xiaodong

Recently a new cover material has been developed for placing on the inside of the greenhouse. This consists of a curtain hung on a wire mesh under the front roof in the night to reduce the heat loss by long wave radiation from the interior at a higher temperature. Initially standard plastic film was used but, since this is transparent to long wave radiation, it can only reduce the heat loss by 20 per cent. The newly developed material is made of PVC or PE film mixed with aluminium powder when the film is manufactured or coated with aluminium by vacuum plating. The insulating efficiency of this aluminium-coated film is much better at between 50 and 70 per cent.

The back and gable wall. The back and gable walls are built of earth or brick with an inner core of insulating materials. The traditional insulation materials are straw and/or dry soil, which are heavy, have a short service life, and are inefficient. Polymer foam, like polystyrene board, is now commonly used owing to its high insulation performance. Photo 4.6 shows a gable wall with polystyrene board between the two brick structures.

Photo 4.6 Greenhouse wall built with polystyrene insulation board
Source: Courtesy of Prof. Guo Xiaodong

Two-span greenhouse. To make more efficient use of land and to reduce material use, another recent development is the two-span greenhouse (see Photo 4.7). This kind of greenhouse also has the benefit of using solar energy more efficiently.

Photo 4.7 Two-span greenhouse
Source: Courtesy of Prof. Guo Xiaodong

CHAPTER 5
Development of rainwater-harvesting-based irrigation systems in Gansu

The challenge of dryland farming and some innovative solutions

The adverse conditions for agriculture experienced on the Loess Plateau in Gansu have already been touched on in Chapter 1. With low mean annual precipitation and high potential evaporation, both surface and subsurface water sources are scarce. The topographical and geological conditions on the plateau, which is criss-crossed with numerous gullies and ravines and comprises easily erodible soil, causing subsidence and frequent slippages, make it unsuitable for any form of water transfer system.

For centuries, local agriculture has completely relied on natural rain. However, this dependence on rain-fed agriculture has been problematic due to low and unreliable rainfall and its uneven distribution throughout the year.

Historically, crop production remained at a very low level. While in an average year the harvest produced a yield equivalent to about 284 kg per person, in dry years the harvest could not even produce enough for the following season's seeds. Since the 1990s, perhaps as a result of climate change, there has been a greater recurrence of drought in the area, causing increasing pressure on all purely rain-fed agricultural food production. Owing to the lack of rain, cash crops such as vegetables, tobacco, fruit, and herbs were not traditionally grown and over 97 per cent of the land was planted with grain crops of low value and thus incomes were very low. Since agriculture has been the primary economic activity in rural Gansu, its unproductive nature has in the past left the population impoverished and underdeveloped. It has long been recognized that any attempt to alleviate poverty in the area would involve improving agricultural productivity. If successful, the implications would be far reaching, not only by helping to lift one of the poorest regions in the country out of poverty, but also for other semi-arid parts of China with significant populations struggling to survive on low-productivity rain-fed agriculture.

Past efforts by Chinese experts to improve conventional dry farming methods have included:

- cultivation methods designed to help the soil absorb as much rainfall as possible, such as deep ploughing, harrowing, and tillage;
- use of straw mulching and plastic sheeting to reduce evaporation from the soil;

- fertilizing crops to increase their resistance to water stress;
- breeding new, more drought-resistant crop varieties.

These measures function in one of three ways: by retaining as much runoff as possible in the soil; reducing evaporation loss; or adapting crops to resist drought. They have been widely adopted and have been shown to be partially effective but their capacity to mitigate the impact of recurrent droughts has been very limited.

In the late 1980s, the area of terraced land amounted to around 1.15 million hectares in Gansu. Terracing the land with inverted slopes can help the soil retain all the rainfall from medium rain events and a large part of rainfall from storms. It is estimated that the soil moisture of terraced land can be between 12 and 25 per cent higher than that of unterraced sloped land. Cultivation (ploughing and tillage) also help to retain rainfall in the soil. However, the soil moisture stored in the summer and autumn through these measures cannot last until sowing takes place in the spring. In this extremely dry and sunny climate (with over 3,000 sunshine hours annually) soil moisture at the time of sowing is usually much lower than required for seed germination.

The water demand of corn conforms better than wheat to the rainfall pattern but there is still a temporal gap between water supply and demand. Especially in mid- and late June, when corn requires more water in its big bell-mouthed period,[7] the soil moisture drops to a low level after the dry spell of spring and natural rainfall often cannot meet demand.

The general principle of conventional rain-fed farming practice is to use soil for moisture storage to meet the water requirement of crops when needed. However, the capacity of the soil to store moisture long term is limited by the size of soil voids and evaporative moisture loss.

Figure 5.1 shows the local pattern of rainfall and crop water demand, and how the time gap between the rainfall and the crop water requirements can be as long as six to seven months. It can be seen from Figure 5.1 that in the

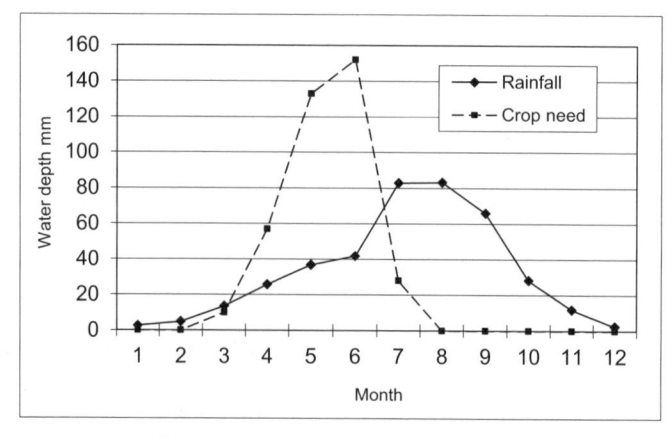

Figure 5.1 Water demand of spring wheat versus natural rainfall in a normal year

three months after seeding in mid-March, rainfall is well below the crops' requirement, while after July there is an excess of rainfall that cannot be used.

In practice the excessive rain in the summer–autumn period cannot be stored in the soil for use by the crop in the following spring since most of the moisture will be lost through evaporation.

According to observations by GRIWAC, soil moisture drops from 14.8 per cent (by soil weight) in autumn to 11.5 per cent in the next spring, corresponding to about 13 mm of water in the 30 cm soil layer. As a result, the soil moisture before seeding in the springtime is often only a little above the wilting point, causing a low germination rate of the seeds and thus a low yield. If there were to be no additional moisture in April to June then the young plants would be subjected to serious water deficit.

Long-term observation shows that the key indicator for a serious drought is the unfavourable rainfall distribution in the year, rather than lower than average annual rainfall. The rainfall in May and June is a particularly critical factor for normal crop growth. During this period, wheat passes the growing stages of jointing,[8] booting,[9] and heading,[10] and corn goes through the stages of seedling to jointing and sprouting. All these stages are critical to the yield of the crops. However, rainfall in these periods only accounts for 19–24 per cent of the annual mean and even less during dry years, representing a major constraint to crop growth. To evaluate the impact of any drought, it is more important to examine the rainfall in May and June than the mean annual rainfall. For example, in 1995, a serious drought occurred on the Loess Plateau of Gansu Province and most of the summer crops withered. Investigations showed that the drought was a 1-in-60-year event, although the rainfall total that year had a recurrence interval of about one in five years. However, the rainfall from May to June in 1995 was 40.9 mm, which has a recurrence interval of 22 years. All the rains in May and June were small events that were ineffective for crops, except for some moderate rains in mid-June. Although significant rainfall did occur in July and August, it had no effect on restoring crop vigour.

After long and careful consideration, the experts and local people finally came to the conclusion that only through irrigation of crops in critical periods could fatal crop damage from water stress be avoided and agricultural productivity increased. The problem was how to get the supplementary water supply the crops required. Since the difficult topographic and geological conditions made any potential water transfer scheme very difficult and costly, the only potentially exploitable water source was rain.

In 1996, based on the successful 1-2-1 Rainwater Catchment Project for domestic water supply, the Gansu provincial government initiated the Rainwater Harvesting Irrigation Project that aimed to extend RWH from a source purely for domestic water supply to providing supplementary water for crops in critical periods. To promote this, GRIWAC and its partners carried out demonstration projects to show the feasibility of supplementary irrigation using only rainwater. The results showed that the role of RWH irrigation in raising the crop yield was significant. According to a recent evaluation of RWH in Gansu, by the end of 2004 there were 2.08 million water tanks built specifically for irrigation. The paved country roads and other existing impermeable surfaces were used as catchments to collect the rainwater. It was estimated that the systems built were supplying supplementary irrigation to an area of about 80,500 ha, including cropland and greenhouses equipped with drip, sprinkler, and pipe irrigation facilities, amounting to 32,000 ha and 6,400 ha, respectively. Although these areas are only a small part of the land area in Gansu Province, they represent a new phase of dry farming development and open the way for further enhancing rain-fed agriculture in the region.

Rainwater-harvesting-based irrigation systems

RWH irrigation systems are composed of the rainwater catchment area, a storage tank, and irrigation facilities.

Existing and purpose-built catchments

To reduce the cost of RWH irrigation systems, the impermeable surfaces of existing structures are often used for the catchment, including paved highways, country roads, threshing floors, and also natural slopes. Among these catchments, the paved highway has the highest collection efficiency

Photo 5.1 Rainwater harvesting for irrigation using a highway for the catchment

and is widely used to lower costs. Photo 5.1 shows a RWH irrigation system using a highway as the catchment.

Experience has shown that a highway paved with either concrete or asphalt typically has a runoff coefficient of 0.6–0.7. Since highways and roads have large areas, water collected from them can feed a number of tanks. The tanks are located down slope from the highway, while the drainage ditch along the edge of the highway can be used as the collection channel. A small temporary dike has to be built in this channel to divert the water to the place of storage. Usually the tanks are arranged in a line parallel to the highway. The land to be irrigated is located below the tank, if the topographic conditions allow, so that irrigation can be achieved using gravity.

When the highway passes a gully, usually a culvert is built for draining both the flood water from the gully and the runoff from the highway. In this case, the water from the culvert can be diverted to the farm by building a dike in the drainage channel of the gully downside of the highway. Photos 5.2 and 5.3 show how Mr Bao, a 61-year old farmer in Zhenggou Chuan village, Qingran township, once one of the poorest villages, diverted water from the highway for his nine water cellars with a total capacity of 430 m^3 to irrigate his two greenhouses and 1 ha of corn and potato. In 2009, his 2-mu (1,334 m^2) cornfield received two water applications from the tank and yielded more than 750 kg/mu. Collecting rainwater from the highway in the past 10 years has made him a fortune of 160,000 CNY ($24,000).

Photo 5.2 The highway passing Bao's house becomes his reliable water source for raising crop yield and improving his life

Photo 5.3 The water collected from the highway is used to irrigate Bao's greenhouses, his field, and the garden

Threshing floors and earthen country roads can also be used as catchment surfaces. Since the rainwater collection efficiency of these types of catchment is low and their area is also limited, the water from them can only be used to fill one or two tanks.

In some situations of extreme water scarcity, a concrete-lined surface is purpose built for catchments on sloping land or hill tops. In these cases, high-value crops such as fruit trees are usually planted to generate sufficient benefits to pay back the building cost. Photo 5.4 shows former wasteland on a hilltop that has been lined with concrete slabs for irrigating an orchard, with 48 tanks built surrounding the catchment area. The cost was relatively high, but was justified and soon paid off due to the high value of fruit (apple and peach) produced by the orchard.

Low-cost greenhouses for harvesting rainwater

Another very efficient catchment surface is the roof of greenhouses. In Gansu many low-cost greenhouses have been built in conjunction with RWH projects. The roof of the greenhouse is constructed using plastic film that is impervious and provides an excellent rainwater collection surface. The runoff

Photo 5.4 Concrete-lined catchment    **Photo 5.5** Greenhouse roof as catchment

coefficient typically exceeds 0.9. The rainwater flows from the roof to a ditch that leads the water to the tank, placed either inside or outside the greenhouse. Photo 5.5 shows the greenhouse and the outside tanks. In semi-arid areas with yearly rainfall of 300–400 mm, rainwater collected just from the roof is not enough for two to three harvests of vegetables, so the ground surface outside the greenhouse is often lined with concrete slabs to provide an additional catchment area, as shown in Photo 5.5.

Rainwater storage tanks for irrigation

Most of the irrigation tanks are the same type as those used for domestic water supply. However, the tanks used for irrigation are generally larger than those used for domestic purposes and are constructed either just by the field to be irrigated or close to the catchment, such as the highway from where the water is conveyed through small drainage channels to the fields. Photos 5.6 and 5.7 show examples of these rainwater irrigation tank arrangements.

For irrigation purposes, sometimes an open surface tank is used, as shown in Photo 5.8. The surface tank has the advantage of a larger storage capacity than the water cellar. It can store water from a large catchment, thus reducing the storage cost. However, under the dry climatic conditions, evaporation losses can be considerable.

Photo 5.6 Tank by the side of cornfield

Photo 5.7 Group of tanks along a canal

Photo 5.8 Surface irrigation tank

Photo 5.9 Water cave for storage

66 EVERY LAST DROP

Photo 5.9 shows another type of storage system used in Gansu: the water cave (local name of *Shuiyao*). These are built in loess cliffs and some of the caves were historically used as homes by people who could not afford to build a house. The tank is located inside the cave so the evaporation loss can be reduced and freezing problems in winter avoided.

Water delivery systems

Where possible, irrigation from the tank is by gravity flow using a siphon pipe system. If the topographic conditions do not allow for this then a hand pump or a mini-electrical pump is normally used for delivering water to the crops. In China a special kind of hand pump designed for lifting water for mini-scale irrigation projects has been developed that can produce pressure of a 15 m-water head, sufficient for a mini-drip irrigation system. Photo 5.10 shows the hand pump used for field irrigation.

The approaches used in the RWH irrigation project are of two types: the modern micro-irrigation system and upgraded traditional methods, as described below.

Photo 5.10 Hand pump for field irrigation

Principles and feasibility of rainwater-harvesting-based irrigation

Compared with conventional irrigation, the RWH systems are micro-scale projects. Consequently, the water that can be supplied for irrigation is very limited. The standard practice in Gansu involves two to three water applications in the crop growing season, one during the sowing stage and one to two applications in the later growing stages. The water available for each application amounts to only 5–15 m^3/mu. Table 5.1 shows the irrigation times and amounts for RWH irrigation systems recommended in the *Technical Standard of Rainwater Harvesting Project* in Gansu formulated by GRIWAC (GBWR and GATS, 1997). The figures in Table 5.1 are only suitable for areas with mean annual rainfall up to 500 mm. For the areas with mean annual rainfall larger than 500 mm, the *Technical Code of Practice for Rainwater Collection, Storage and Utilization* (China Ministry of Water Resources, 2001) recommends the irrigation quotas shown in Table 5.2.

Table 5.1 Irrigation frequency and amounts for rainwater harvesting systems

			Staple cereal crop		Fruit trees	Vegetable
			Summer crop	Autumn crop		
Irrigation frequency in growing season	Mean annual rainfall	300 mm	3–4	3–4	4–5	8–9
		400 mm	2–3	2–3	3–4	6–8
		500 mm	2–3	1–2	2–3	5–6
Application amount m^3/ha	Drip irrigation on plastic film		150–225	150–225	120–225	150–225
	Manual irrigation, injection irrigation		75–150	75–150	75–120	75–150

Table 5.2 Application frequency and quota for rainwater harvesting irrigation

Crop	Irrigation method	Irrigation frequency under mean annual precipitation (mm)		Quota of each application, m^3/ha
		250–500	≥500	
Corn, wheat	Irrigation with seeding	1	1	45–75
	Manual irrigation	2–3	2–3	75–90
	Irrigation on plastic film	1–2	1–2	45–90
	Root injection	2–3	1–2	30–60
	Furrow under plastic film	1–2	2–3	150–225
Vegetables	Drip	5–8	6–10	120–180
	Micro-sprinkler	5–8	6–10	150–180
	Manual irrigation	5–8	8–12	75–90
Fruit trees	Drip	2–5	3–6	120–150
	Bubble	2–5	3–6	150–225
	Micro-sprinkler	2–5	3–6	150–180
	Manual irrigation	2–5	3–6	150–180
Paddy	Water-saving cultivation		6–9	300–450

We can see that the irrigation amounts listed in both Tables 5.1 and 5.2 are much less than for conventional irrigation. Irrigation using RWH is a special kind of water-saving irrigation approach that we can term 'low-rate irrigation' (LORI). When GRIWAC first carried out trials and demonstrations with RWH irrigation, many people doubted the effect on crop yields of irrigating with such small amounts of water. Earlier experiments and demonstrations using RWH irrigation had shown that although the water used is very little, the crop yield increase was remarkable when compared with rain-fed crops. According to investigations by the Gansu Bureau of Water Resources, the crop yield with RWH irrigation in dry farming land could be raised by 40 per cent on average. Table 5.3 is a summary of the results in the testing and demonstration carried out by GRIWAC, Gansu Academy of Agricultural Sciences, and the Gansu Agricultural University. In Table 5.3, water supply efficiency (WSE) is the yield increase divided by the total irrigation water in the crop growing period. Results shown indicate the high productivity of the irrigation water and the potential yield increase by adopting LORI with the RWH system.

The reason why crop yields can be raised significantly with such small amounts of water can be explained as follows. Water supplied using LORI methods in conjunction with RWH provides only 10–15 per cent of the crop water needs for the whole growing season. Most of the crop water demand is still met from natural rain. Nevertheless, with the water supplied only during the critical periods the crop can avoid permanent damage due to water stress, thus enabling the crop to effectively utilize natural rainfall later in the season. The crop would not survive or would be seriously damaged if no water was supplied to it in the critical period, and any rain later in the season would be useless. Therefore the role of LORI is actually to raise the water use efficiency (WUE) of the natural rain, the main water supplier to the crop. This assumption is confirmed by the marked difference in WUE, with and without LORI. Here the WUE equals the yield divided by the total crop water consumption, including the effective natural rain, the soil moisture, and the supplemental irrigation amount. Detailed investigations to be discussed later

Table 5.3 Results from testing and demonstration projects

Crop name	Irrigation amount (m^3/ha)	Yield (kg/ha)	Yield increase (%)	WSE (kg/m^3)
Spring wheat	225–300	1,990–6,843	10.5–88.3	1.65–3.9
Corn	375–405	2,940–9,050	19.6–88.4	3.11–5.7
Potato	405	27,696	30.6	10.95
Millet	300	2,583–2,750	20.5	1–1.62
Broom corn millet	300	4,011–4,258	6.8–13.4	1.5–1.55
Oil sunflower	450	2,626–3,000	19.8–65	1.65–3.45
Linseed	225	1,590–2,505	44.7–120.6	3.03–6.08

Source: Courtesy of GRIWAC et al. (2002)

show that WUE with LORI increased by 29–59 per cent for wheat and 15–35 per cent for corn, when compared to purely rain-fed crops. It should be noted that the soil moisture and the irrigation water amount also originate directly from rain in the locality. This is the rain stored in the soil post harvest and the rain stored in the RWH system. The results indicate that although LORI water only comprises 10–15 per cent of total crop consumption, by adding this small percentage at a critical time, the overall efficiency of the total water use is significantly raised.

How then is LORI carried out to make it so efficient? According to the experiences in Gansu the following three approaches should be followed:

- The principle of deficit irrigation (limited irrigation) should be applied, as opposed to the sufficient irrigation approach that promotes maximum yield by fully meeting the water demand of the crop at every growing stage. Deficit irrigation aims to achieve the maximum WSE/WUE. When the available water amount is unlimited and the land resources are the restricting factor then the sufficient irrigation approach would be preferable to get maximum benefit from limited land. However, when the available water is limited, then the deficit irrigation approach should be applied. The areas where RWH is adopted usually have serious water shortages and belong in the latter situation. The limited amount of rainwater stored in the tank should be used very carefully by only applying it at critical periods of crop growth and by only partially meeting the crop's water requirement. The so-called critical periods are the periods when the crop is subjected to serious water stress and the damage to the crop would be irrecoverable, even with watering or natural rain at a later stage. One of the main focuses of the RWH research in Gansu has been to find the critical periods for different crops.
- Water saving irrigation methods need to be adopted to get the highest efficiency from any water application. However, of the two methods commonly used with RWH, modern micro-irrigation and traditional simple methods, the latter are still used by most people due to their affordability and high irrigation efficiency.
- Water applications should target the root zone of the crop to avoid evaporation from the soil surface as far as possible. To quote the farmers, 'we are irrigating the crop, not the land'. The widespread use of plastic sheeting also helps enhance water conservation.

CHAPTER 6
Irrigation methods using rainwater

The irrigation methods using RWH can be divided into two types: modern micro-irrigation, including drip, mini-spray, and bubble irrigation[11] and simpler, lower-cost methods.

Simple and affordable irrigation methods

In Gansu, RWH projects have been carried out in many impoverished rural areas where most of the farmers could not afford to buy modern irrigation equipment. In response, local technicians and farmers have adopted and developed many simple and affordable irrigation methods, some of which are outlined below.

Irrigation during seeding

This method has been used in the dryland farming areas of China for many years to ensure the emergence of the seedlings. In this method a small amount of water is poured in the hole where seed is to be dropped immediately before sowing. After the watering and sowing are finished, plastic film is used to partially cover the land. The operation involves: digging holes or ditches, pouring water, sowing, applying fertilizer, refilling and compacting the soil, laying plastic sheeting, and covering the edges of plastic with soil. In China the operation is done either manually or by machine. There are many kinds

Photo 6.1 Integrated machine for sowing and then laying plastic sheeting

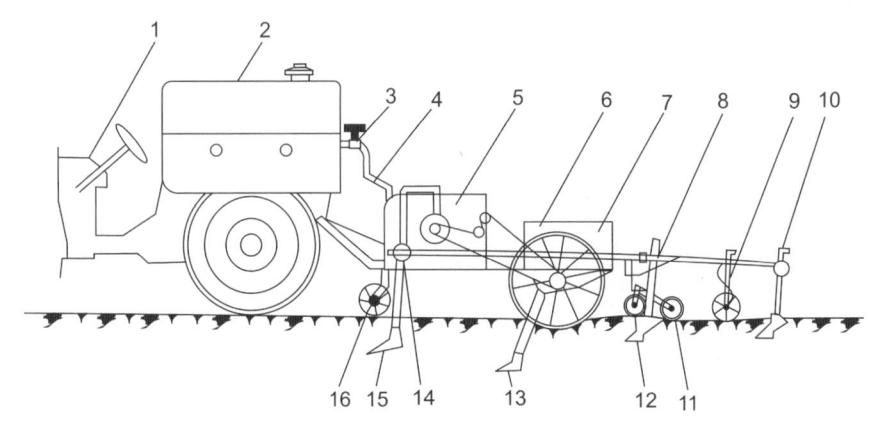

Figure 6.1 Cross-section of multi-function sowing machine
Note: 1 Tractor, 2 Water tank, 3 Valve, 4 Water supply hose, 5 Second water tank, 6 Fertilizer container, 7 Seed container, 8 Frame for plastic film sheeting, 9 Wheel to press film, 10 Shovel for covering soil on the film, 11 Plastic film unrolling wheel, 12 Plough for ditch to lay the film, 13 Plough to dig ditch for seeding and applying fertilizer, 14 Plug in water outlet, 15 Plough to dig ditch for water application, 16 Wheel to control the depth of seeding.

of sowing machines with integrated functions for this operation. Photo 6.1 shows an integrated machine for sowing and laying plastic sheeting and Figure 6.1 shows the components of a multi-function sowing machine.

As a rough estimate, the irrigation water requirement for corn at sowing can be taken as between 45 and 75 m^3/ha (Table 5.2). When irrigating corn in Gansu, about 1 litre is applied per seed hole. Typically there are 45,000–52,500 plants per hectare so the water requirement is about 45–52.5 m^3/ha. Observations in Inner Mongolia show that with this amount of water application, the wetted diameter and depth in the soil is 25 cm, while the wetted percentage of the field is about 25 per cent. Since the moisture is confined to the area surrounding the seeds, this small amount is equivalent to 15–20 mm rainfall. When the operation is done by machine, the amount of water and fertilizer applied to the seeds can be adjusted. Water is applied to a point a little deeper than the seeds to link it with the existing soil moisture below.

Seeding is the most critical period for irrigation in order to raise the crop yield. If the soil is dry at this time, a significant percentage of seedlings will not emerge, causing a major loss in yield. It has often been observed that irrigation with a small amount of water during seeding can even double yield. Photo 6.2 shows a comparison of crop performance with and without irrigation during seeding in Inner Mongolia in 2003. Table 6.1 lists results of yield increases with irrigation during seeding. It can be seen that in a medium-dry year yield can be increased by 30–40 per cent, while in a dry year yield may be increased by 50–70 per cent, even up to 143 per cent.

IRRIGATION METHODS USING RAINWATER 73

(a) (b)

Photo 6.2 Comparison between (a) with and (b) without irrigation during seeding

Table 6.1 Effect of irrigation during seeding on yield of corn

Location	Year	Moisture status	Yield with irrigation (kg/ha)	Yield without irrigation (kg/ha)	Yield increase (kg/ha)	Increase (%)
Zhaodong Municipality	1988	Moist	5,700	4,500	1,200	27
	1989	Medium	5,100	4,000	1,100	30.3
Zalandun Municipality	1992	Dry	7,600	3,100	4,500	143
Tuoquan County	1990	Medium	6,000	4,300	1,700	37.5
Keyouqian County	1994	Dry	5,400	3,100	2,300	74.7
		Dry	5,200	3,100	2,100	68.4
	1995	Dry	8,100	5,400	2,700	50.0
		Dry	9,300	5,400	3,900	73.4

The increased productivity with irrigation is very high. Table 6.2 shows results of testing by the Water Resources Institute of Inner Mongolia Autonomous Region. It shows that the yield increase of corn per unit water amount can be as high as 19.6–37.6 kg/m^3.

Table 6.2 Water productivity of irrigation during seeding of corn

Treatment	Year	Irrigation amount (m^3/mu)	Yield (kg/mu)	Yield increase (kg/mu)	Yield increase (%)	WSE (kg/m^3)
No irrigation	2003	0	413.8			
	2004	0	365.6			
	2005	0	385.2			
Irrigation during seeding	2003	3	506.0	92.2	22.3	30.7
	2004	3	478.3	112.8	30.8	37.6
	2005	3	444.1	58.9	15.3	19.6

Irrigation using plastic film

In north and northwest China plastic sheeting is widely used to avoid evaporation loss and to increase soil temperature. Tests have shown that when a field is covered with plastic film the evapotranspiration from the crop can be reduced by 30 per cent and the ground temperature in springtime raised by 2–3°C. In this area, corn has a much higher yield than wheat because it has a longer growing period, allowing greater exploitation of available rainfall. However, in mountainous areas at altitudes higher than 2,000 m, early frosts often occur before corn reaches maturity. If plastic sheeting is used the corn can be sowed two weeks earlier, giving it more time to reach maturity. Since the adoption of plastic sheeting, corn can now also be planted in more mountainous areas. Another advantage of plastic sheeting is that most of the weeds under the film die due to lack of air so the need for weeding is greatly reduced.

The plastic film is placed after sowing, watering, and fertilizer application. Initially the germinating seedling is protected from frost and desiccation by the plastic film, which effectively creates a mini-greenhouse. When the seedling appears, the film is manually punctured in a cross shape to enable the young plants to emerge. Later the holes are enlarged to let the natural rain and irrigation water reach the soil. The holes are placed at points of depression. The land should be well prepared with a flat and smooth surface to avoid water retention. Water is applied manually, using a bucket or a hose to divert water

Photo 6.3 Watering crops through holes in the plastic film

from the tank, so irrigation efficiency is very high. Irrigation water amounts to about 1–3 kg for each plant and 45–90 m^3 per hectare. Photo 6.3 shows how farmers irrigate crops through the holes of the plastic film.

Furrow irrigation using plastic film

Furrow irrigation is another kind of RWH irrigation practice using plastic sheeting. Two lines of ridges are planted with a ditch in between. Another ditch dividing each pair of rows is covered with plastic film, which also covers the slopes of the ridges. The bottom of the furrow between the rows remains open so water (rain and irrigation water) can be absorbed by the soil. Figure 6.2 illustrates this kind of irrigation.

Furrow irrigation saves water loss from evaporation and tests have shown the irrigation water requirement can be reduced by more than 60 per cent compared to conventional surface irrigation methods.

Injection irrigation

Injection irrigation supplies water to the crop by injecting it directly into the root zone. Farmers in the Ningxia Autonomous Region first used the fertilizer/pesticide injector for injection irrigation. See Figure 6.3. Water is injected into the crop root zone with a manually driven compressor.

Later some specially designed water injectors were developed. The injector is linked to a tank with a hose. A mini-size submersion pump is used to provide water pressure. If the tank is at a higher location than the irrigated field, the water can flow by gravity and a pump is not required. Sometimes, one tank can provide water for several injectors.

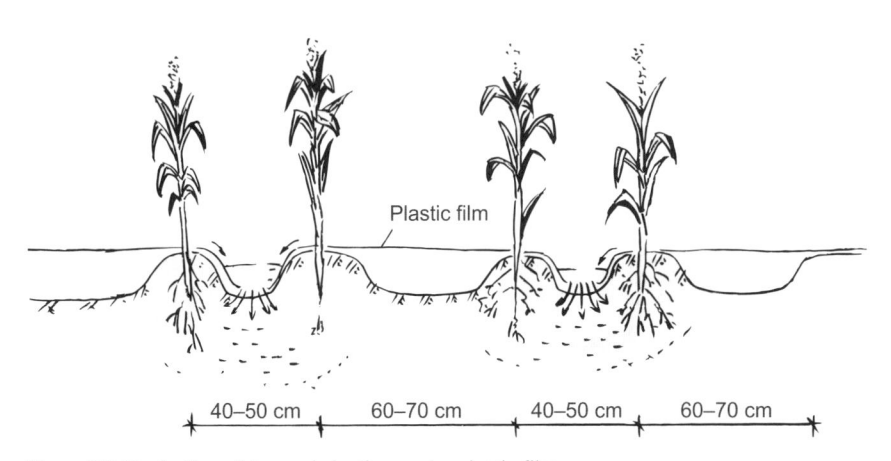

Figure 6.2 Illustration of furrow irrigation under plastic film

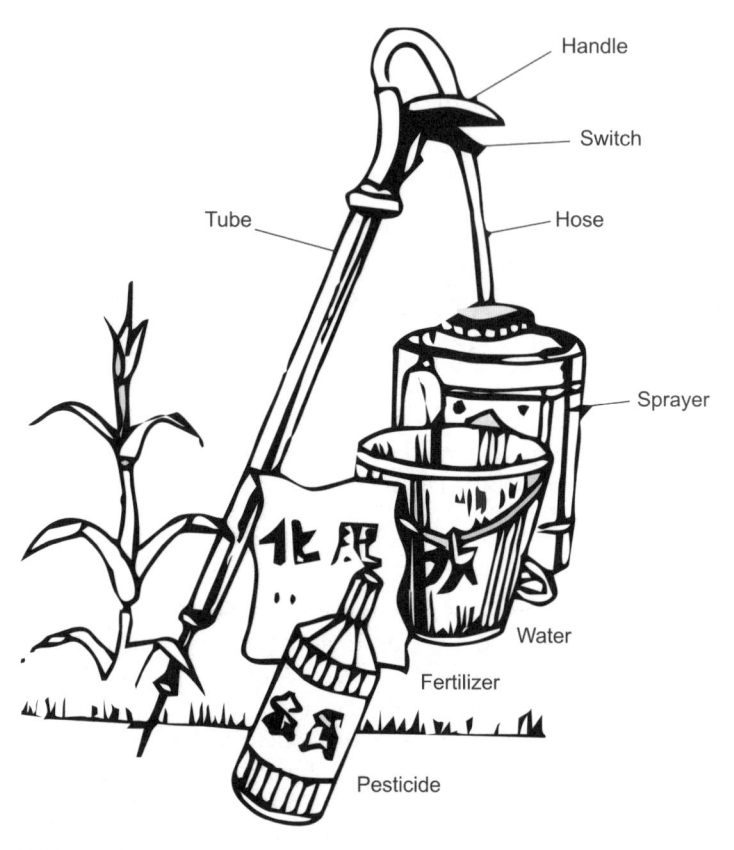

Figure 6.3 Water injector

Advantages of injection irrigation are:

- Water, fertilizer, and pesticide can be applied in one operation.
- The ground surface is kept dry during irrigation so moisture evaporation from the soil is minimized.
- The water requirement is very low. Each application only needs about 1–1.5 kg of water.

For 1 ha of corn, for example, with a crop density of 45,000 plants per hectare only about 45–67 m^3 of water is needed. The disadvantage of the method is that it is highly labour intensive.

Injection irrigation is best suited for low-density plants, such as fruit trees, melon, grapes, and corn. It has been used for melons in Ningxia Province with six to eight applications for the whole growing season and using about 30 m^3/ha each time. The yield of melons was more than 30 t/ha. The utilization of a total of 300 m^3/ha of water over five applications for corn yielded 5.3 t/ha.

Seepage irrigation using a porous pot

Seepage irrigation using a clay vessel with porous walls has been adopted in Luoma village in Gansu for fruit tree and corn irrigation. Figure 6.4 illustrates seepage irrigation for trees and corn.

The unglazed vessel is filled with water, which is absorbed gradually by the soil through capillary action. Water loss due to evaporation and deep seepage is avoided. The labour needed for water application using vessels is less than that for injection irrigation. However, this kind of seepage irrigation is more suitable for trees than annual crops. Placing the vessels after seeding and collecting them after harvest requires a lot of labour and vessels can easily be broken during the process.

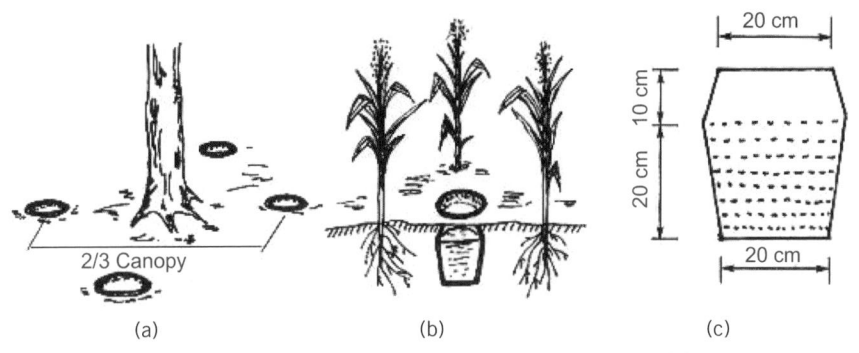

Figure 6.4 Seepage irrigation for fruit trees and corn using porous pots: (a) four vessels around a tree; (b) three corn plants per vessel; and (c) a percolation vessel

Photo 6.4 Manual irrigation using a hose

Manual irrigation

In addition to irrigation during seeding, manual irrigation is the most popular method for RWH irrigation in China because of its low cost and simplicity. Water is poured on the crop root zone manually, using either a bucket or a hose from a tank. Where a hose is used, the flow is pumped. If the tank is located above the field, water can be first siphoned from the tank and then gravity fed. The field is often covered with the plastic film and water passes through the holes in the film. The amount of water required for manual irrigation ranges from 75–90 m^3/ha, depending on the crop type and the experience of the water user.

Micro-irrigation systems

In Gansu, two types of micro-irrigation systems are commonly used in conjunction with RWH: drip irrigation and small sprinkler units.

Drip irrigation system

The drip system consists of the water source, a pivot, and a pipe system. In Gansu the water source is usually rainwater from a water cellar (*Shuijiao*) with a storage capacity ranging between 30 and 80 m^3. The function of the pivot is to take water from the tank and meet the required flow rate, water quality, and water pressure for pressurized irrigation. It usually consists of a pump, filter, chemical container (optional), and the controlling and measuring equipment (pressure meter, flow meter, and valve). The pipe system consists of the main

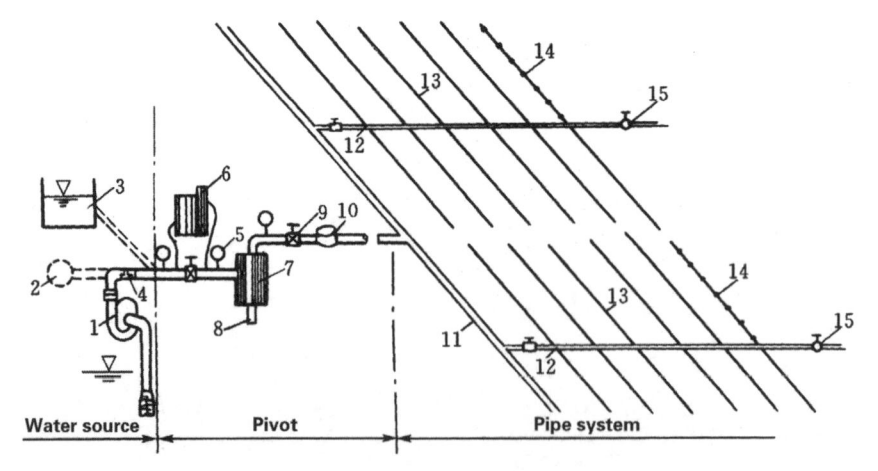

Figure 6.5 Illustration of mini-drip irrigation system
Note: 1 Pump, 2 Water supply pipe (optional), 3 High location tank (optional), 4 Non-return valve (optional), 5 Pressure meter, 6 Chemical container (optional), 7 Filter, 8 Sludge drainage, 9 Valve, 10 Flow meter, 11 Main pipe, 12 Branch pipe, 13 Lateral pipe, 14 Emitter, 15 Flush valve

pipe, branch pipes, and laterals. Since the drip systems used in conjunction with water cellars are very small scale, branch pipes are not usually necessary. The laterals allow for the uniform distribution of water to the field. The outflow from the pipe is through a device called an irrigator or emitter. The arrangement of a typical drip system is illustrated in Figure 6.5.

The pump is either a mini-submerged electrical pump or a hand pump. The flow rate and lift head of the submerged pump is 3–4 m^3/hr and 10–15 m, respectively.

It is essential to prevent the mini-irrigation system from clogging a filter. In general, the filter system includes a screen filter, sand-gravel filter, and rotating silt separator. With RWH, the mini-irrigation system is very small, so the screen-type filter for drip systems is made of plastic with 120# mesh screen (120 openings on every square inch of the screen). Figure 6.6 illustrates the main components of a screen filter.

Drip systems have the advantage of applying water, fertilizers, and pesticides directly to the root zone in one operation. These should be in liquid form in a separate container and applied to the crop together with the water. There are two kinds of applicator: Venturi and pressure differential. When the water flows through the Venturi pipe with a narrower section, a vacuum occurs and the fertilizer solution is sucked into the pipe. In the pressure differential applicator, the fertilizer solution is driven by the pressure difference built up in the pipe.

There are many kinds of drip emitter, but in Gansu's RWH irrigation systems, a drip line system with a built-in emitter is widely used. The emitter is installed in the inner side of the pipe during manufacture. An example is shown in Photo 6.5. It has a number of advantages including a uniform

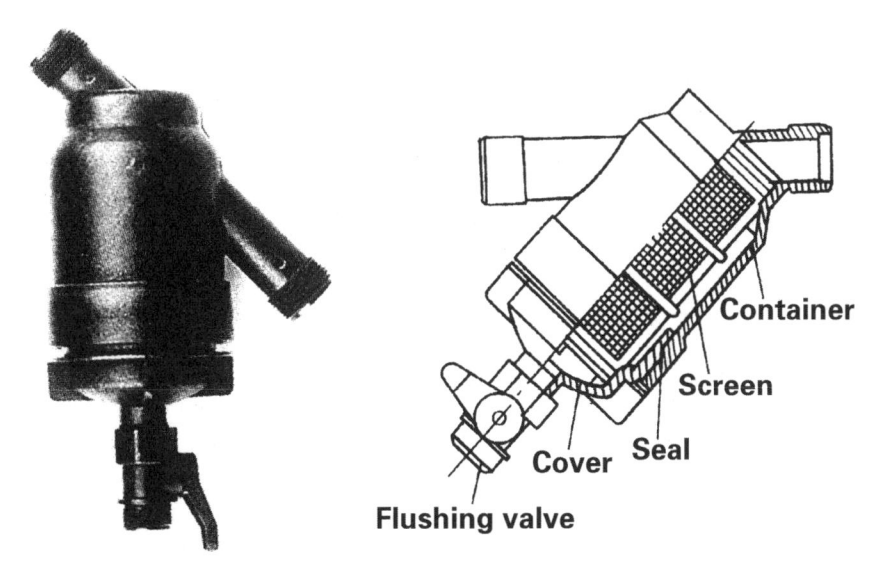

Figure 6.6 Screen-type filter

80 EVERY LAST DROP

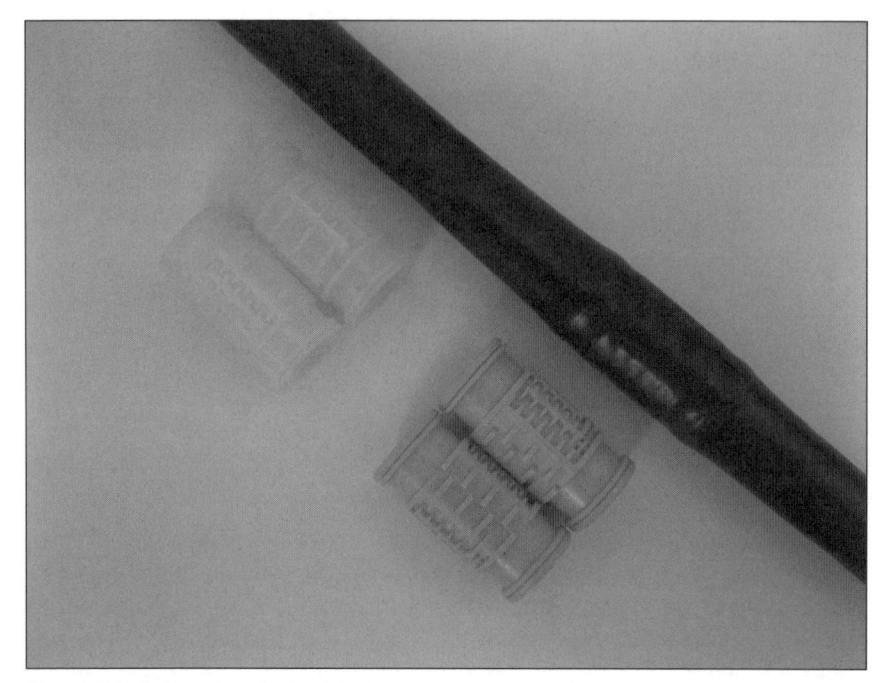

Photo 6.5 Built-in emitter in the drip line

outflow, anti-clogging feature, and is both affordable and easy to install. It is widely used both in greenhouses and field crop situations. The service life is more than five years for pipe systems with a wall thickness of 0.6–0.8 mm, and one to three years for a wall thickness of 0.2–0.4 mm. The specification of drip line pipes produced in China is listed in Table 6.3.

There are two layouts for drip irrigation systems used in conjunction with RWH: one for field crops and one for greenhouse use.

For field crops, the drip line moistens a strip of land 0.6–0.8 m wide in sandy soil and 0.8–1.0 m in clay soil. The drip line is placed parallel to the crop line at intervals equal to the width of the moistened strip. The drip system

Table 6.3 Specification of drip line produced in China

Outer diameter (mm)	Wall thickness (mm)	Working pressure (kPa)	Flow rate (l/s)	Interval of emitter (m)	Price (2001) (CNY/m)
16	0.6	100	2.8	0.3, 0.4, 0.5	1.0
16	0.4	100	2.8	0.3, 0.4	0.8
16	0.2	100	2.8	0.4	0.4
12	0.4	100	2.7	0.3, 0.4	0.6

Note: 8.3 CNY = US$1

supplied from rainwater tanks should be used on flat land where available or on terraced fields in hilly areas, so the lateral is able to irrigate both sides of the crop. Since the stored volume is small, the maximum area that one tank with a volume of 30–60 m³ can provide is 2–4 mu (1,334–2,667 m²). To reduce costs, after irrigation from one tank is complete, the whole drip system including the pump, filter, and valve, as well as all the piping (main pipe, laterals, and emitters) should be moved from one tank to another. Only when irrigating orchards is the drip system fixed. In some villages, there are households specializing in the operation and maintenance of drip irrigation equipment. They own a system and provide a service to other households in return for payment. This form of irrigation management is highly cost efficient as most households avoid the initial cost.

The following example is for a movable drip system for field crop of 2 mu (,1334 m²) as recommended by GRIWAC. This system works with two tanks of 30 m³ each (or one tank of 60 m³ capacity). An RB 1.5-type hand pump that can supply a flow rate of 1.5 m³/hr with a water head of 15 m is used for delivery of the water. The filter is a 120# mesh screen filter with inlet and outlet diameters of 2.5 cm (1 inch). The laterals are made of PE black piping with a thickness of 0.6 mm and an outer diameter of 16 mm. The emitters with a total flow rate of 2.8 m³/hr are placed on the drip line at intervals of 0.4 m. The system includes four laterals, each with length of 30–40 m, depending on the shape and size of the land. The length of the branch pipe is about 40 m. Figure 6.7 illustrates the layout of the drip system. Table 6.4 lists the main equipment and materials of the system and cost (in 2001) for reference.

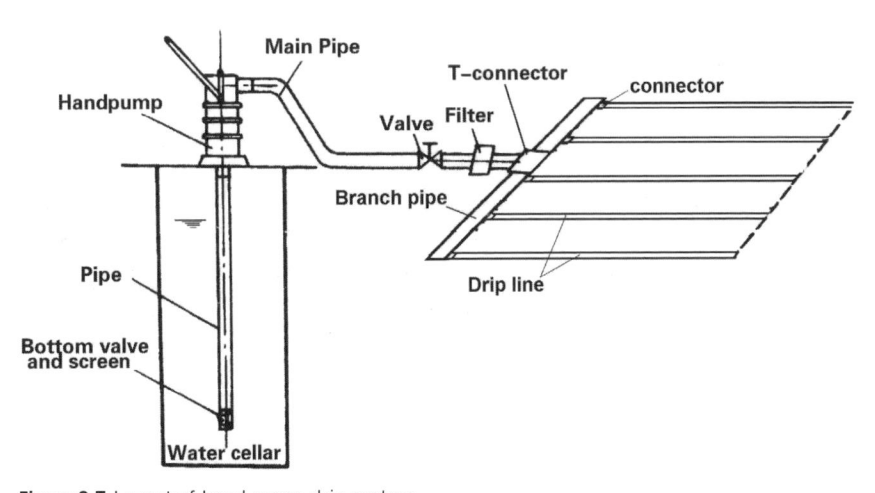

Figure 6.7 Layout of hand pump drip system

Table 6.4 Equipment for movable hand pump drip system for 2 mu of field crop

Name	Specification	Quantity	Unit price (CNY, 2001)	Cost (CNY, 2001)
Hand pump	RB1.5	1	150	150
Screen filter	1 inch	1	66	66
PE pipe	φ25	40 m	2.3	92
Drip line	0.4	160	0.8	128
Perforator	φ14	1	10	10
Fittings and other				30
Total				476

Note: 8.3 CNY = US$1

The hand pump can be replaced with an electric submersible pump. Because of the larger flow rate of the pump, it can serve up to four tanks with a volume capacity of 30 m³ each, or two tanks with capacity of 60 m³. The quantity of the drip line in Table 6.5 should be doubled and the system can irrigate a field area of 4 mu. According to hydraulic calculations, the total water head of the system is 18 m and the working head of the emitter is 10 m. The flow discharge of the whole system is 3.6 m³/hr. The submersible pump QDX4-20-0.55 with flow capacity of 4 m³/hr and lift head of 20 m is used in this situation.

All the drip lines are put under a plastic film to avoid evaporation. Photo 6.6 shows laying of the plastic-sheeted drip lines by machine.

Photo 6.6 Drip line and plastic sheeting can be laid by machine

IRRIGATION METHODS USING RAINWATER   83

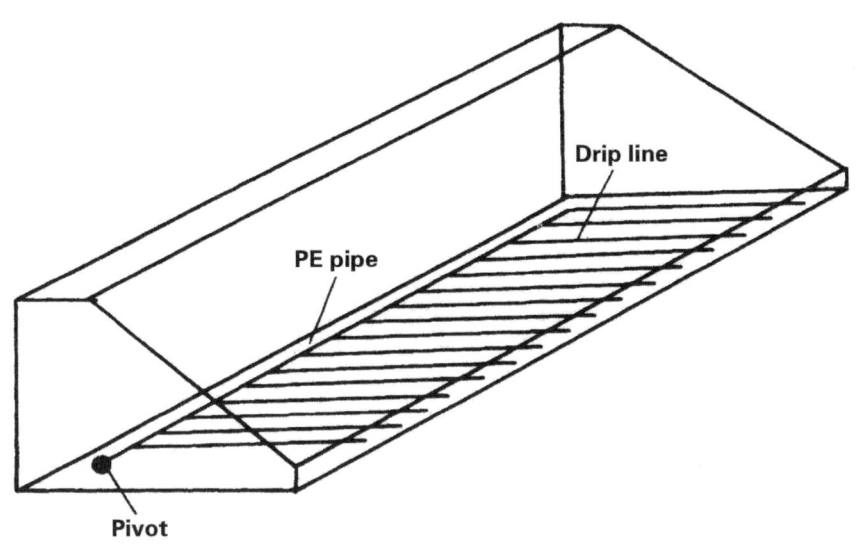

Figure 6.8 Layout of fixed drip system for 400 m² greenhouse

Another example is the fixed drip system for greenhouses. In Gansu, greenhouses are usually rectangular with the longest side arranged in an east–west direction. The width of the greenhouse is typically 6–8 m. The following example is for a 0.6 mu (400 m²) greenhouse with a length of 60 m and width of 7 m. The branch pipe is placed along the longer side and the laterals are arranged in a north–south direction. The lateral is laid under a plastic film with intervals of 1 m for every two rows of vegetables. Emitters are put every 0.3 m along the lateral. The branch pipe is PE black pipe with an outer diameter of 32 mm and a length of 65 m. The equipment used in the system is listed in Table 6.5 and illustrated in Figure 6.7.

Table 6.5 Equipment for movable electric pump drip system for a 4 mu field crop

Name	Specification	Quantity	Unit price (CNY, 2001)	Cost (CNY, 2001)
Submersible electric pump	QDX4-20-0.55	1	300	300
Screen filter	1 inch	1	66	66
PE pipe	ϕ32	55 m	2.9	159.5
Drip line	0.3	385	0.8	308
Fertilizer container	10 litre	1	135	135
Valve	Dg25	1	38	38
Perforator	ϕ14	1	10	10
T-connector	ϕ15	55	0.9	49.5
Plug	ϕ15	55	0.7	38.5
Others				28.3
Total				1,132.8

Note: 8.3 CNY = US$1

The electric pump specified in Table 6.5 can also be replaced by a hand pump that can create enough pressure for a drip system in the greenhouse. In this case, a water tank in or alongside the greenhouse is installed at a height of 2.5 m above the ground as a buffer to ensure a constant water supply to the drip system.

Small sprinkler unit

The small sprinkler unit is mainly for mitigation during drought conditions. It can be moved from field to field easily for crop and orchard irrigation. This unit can be owned by several households or by one family responsible for irrigation. Photo 6.7 illustrates a sprinkling unit.

The unit is composed of a gasoline engine pump set, pipes, hoses, and sprinklers. The 1.18 kW gasoline engine consumes fuel at 0.8 litres per hour. The pump is a self-suction pump with a lift head of 25 m and outflow of 6 m^3/hr. The maximum number of sprinklers is eight and each sprinkler has a flow rate of 0.6 m^3/hr. The irrigation area and the sprinkler intensity versus number of sprinklers are shown in Table 6.6.

Photo 6.7 Sprinkling unit with rainwater tank as source

Table 6.6 Irrigation area and sprinkler intensity versus number of sprinklers

Number of sprinklers	8	7	6	5	4	3
Irrigation area (ha)	0.064	0.056	0.048	0.04	0.032	0.024
Sprinkler intensity (mm/hr)	9.4	10.7	12.5	15	18.74	25

Micro-catchments using rainwater concentration
Plastic sheeting laid during planting

The micro-catchment is a cultivation practice frequently adopted by farmers, usually for common crops such as corn, potato, millet, etc. It is not an irrigation practice but a method of rainfall concentration using plastic sheeting. Natural rainfall on the non-planted area is redirected to the cropping area to increase the rainwater that can be used by the crop. To increase the concentrating effect, the non-planted area is usually lined with plastic film with a thickness of 0.01 mm. The micro-catchment field is prepared by alternating the ridges and furrows spaced at 30–40 cm. The crop is planted in the furrow and the ridge is covered with plastic film. The ratio of the width of the ridge to that of the furrow indicates the extent of the rainwater concentration for the crop.

Test results and calculations for 10 rural locations, under normal rainfall conditions, revealed the total water deficit during the growing period for the four main crops as shown in Table 6.7.

To design the micro-catchment to meet the water deficit we need to determine the ratio of the width of the ridge to that of the furrow using the following equation:

$$r = \frac{kP_d}{EP_n} - 1 \tag{6.1}$$

where r is the ratio of the width of ridge to furrow, P_d is the total crop water demand in millimetres, P_n is the annual natural rainfall during the growing period, E is the rainwater collection efficiency of plastic film, and k is a factor that represents the percentage of the water demand being met on the basis of the whole crop-growing period. The purpose of Equation 6.1 is to quantify how to compensate for the water deficit through rainwater concentration. In Equation 6.1, as a rough estimate with built-in safety factor, E is usually taken as 0.9. The factor k is determined with economic considerations in mind. A larger factor of k demands a wider ridge (a larger r), which means less land can be planted. The yield increase by concentrating more rainfall to the planted area would not be able to pay back the decrease of yield due to loss of planting area and the cost of plastic film. If k is too small, then little or even no rainwater concentration exists. There should be an optimum value of factor

Table 6.7 Water deficit in the growing period for the main crops

Crop name	Growing period	Total water demand (mm)	Rainfall during growing period (mm)	Water deficit (mm)
Spring wheat	15 March–25 July	347	196.2	150.8
Corn	15 April–31 August	436	302.8	133.2
Millet	10 May– 20 August	316	245.7	70.3
Linseed	25 April–10 August	298	233.6	64.4

Table 6.8 Results of *r* versus mean annual rainfall as recommended by GRIWAC

Mean annual rainfall (mm)	300–350	350–400	400–450
Wheat	0.6–0.7	0.4–0.6	≤0.4
Corn	0.3–0.4	0.2–0.3	≤0.2

Source: Courtesy of GRIWAC et al. (2002)

k to get the highest yield under particular rainfall and land conditions. From Equation 6.1 it can be seen that value *r* will decrease with any increase of the rainfall in the growing period. Table 6.8 shows the results of *r* for different mean annual rainfall totals as recommended by GRIWAC. Here *r* for corn is smaller than for wheat because the water deficit for the former is less than that for the latter.

Photo 6.8 shows micro-catchments for wheat and corn.

Micro-catchments can also be used for planting trees. One way is to cover the land surrounding the trees with plastic film and another way is to plant trees in deep pits to concentrate rainwater to the root zone of the tree. Both practices give good results in ensuring a high rate of survival for young plants. Photo 6.9 shows micro-catchments for tree planting with plastic sheeting and using deep pits.

Photo 6.8 Rainwater concentration using plastic sheeting

Photo 6.9 Micro-catchments for tree planting

Table 6.9 Yield increases by adopting micro-catchments in corn and potato tests

Treatment	Corn		Potato	
	Yield (kg/ha)	Yield increase (%)	Yield (kg/ha)	Yield increase (%)
Rain-fed	3,530		23,313	
Plastic sheeting	4,302	21.9		
Micro-catchment	4,619	30.8	27,695	18.8

Source: Courtesy of GRIWAC et al. (2002)

Tests of micro-catchments for corn and potato show that the yield can be raised significantly. The tests for corn compared the yield between rain-fed only, rain-fed with plastic sheeting, and the micro-catchment. The tests for potato compared the yield between rain-fed without plastic sheeting and micro-catchment. The results are shown in Table 6.9.

The purpose of micro-catchments is to compensate for the water deficit by concentrating rainwater runoff from the non-planted area onto the crops. This mitigates water stress and increases yields by 18–30 per cent, as shown in Table 6.9. The technique is very popular with local farmers due to its simplicity and low cost.

It should also be pointed out that the use of plastic sheeting for micro-catchments has not only brought about benefits in Gansu, but also caused some serious waste management problems. This has been due to the large amount of plastic waste generated, which is usually found in the fields after harvest, causing so-called 'white pollution'. A degradable film has been developed in Gansu but the transparency of this film is worse than the plastic one and the price is higher. Further improvements and developments are therefore urgently needed.

Plastic sheeting used after harvest

Harvesting of the main crops is normally finished by the end of August. However, rainfall in September and October on average amounts to about 22 per cent of the annual total. Much of the soil moisture resulting from rainfall in this period is lost before seeding the next spring. GRIWAC has therefore tested the effect of moisture conservation using plastic sheeting on fields after harvest at 10 locations. The results of this study are shown in Table 6.10.

Table 6.10 Effect of plastic sheeting after harvest on soil moisture conservation

Water content after harvest (%)	With plastic		Bare field		Moisture conservation by plastic sheeting (mm)
	Water content next spring (%)	Moisture loss in top 30 cm soil (mm)	Water content next spring (%)	Moisture loss in top 30 cm soil (mm)	
14.8	13.8	3.8	11.5	11.8	8

Source: Data from GRIWAC et al. (2002)

It can be seen that the soil moisture during the springtime can be increased by 8 mm in the top 30 cm soil layer, equivalent to a water application of 75 m³/ha, which is very beneficial for seed germination. The wilting point of the crops is about 8 per cent by weight. The moisture content of the bare land is closer to the wilting point in the springtime when seeding is carried out.

Optimizing the irrigation schedule

Optimizing the irrigation schedule involves the formulation of three parameters: the water amount, the timing, and the frequency of applications. A wise selection of these parameters can significantly raise the crop yield and the WUE. Since this form of irrigation relies on a finite water supply, the limited water stored in the tank is of great value to the water-deprived population and has to be used very carefully. It is therefore particularly important to have a good irrigation schedule. GRIWAC and its partners have carried out a number of tests over the years to find out the best irrigation schedule for different crops in the area. The following is a brief summary of the results.

In the Loess Plateau of Gansu Province, the staple crops are spring wheat and corn. Vegetables were not commonly planted before implementation of the RWH project but have now become more popular and offer the best prospect to improve the livelihoods of farming families due to the income they can generate.

Spring wheat

On the Loess Plateau of Gansu Province, spring wheat is sown around 15–20 March and harvested at the end of July. The total growing season lasts for about 132 days. Tests have shown annual water demand is about 3,500 m³/ha (equivalent to 350 mm of rainfall). Natural rainfall in the growing season is about 200 mm (2,000 m³/ha) leaving a water deficit of about 1,500 m³/ha, which occurs mainly in May and in the first half of June. Irrigation in any growing stage enhances the yield but the productivity of water applied varies widely depending on the timing of its application.

GRIWAC and its partners including the Gansu Academy of Agriculture Sciences and Gansu Agriculture University carried out a number of tests in the period 1997–9. Table 6.11 shows the comparison of yield using different irrigation water amounts in the growing season.

From Table 6.11 we can see that in water-short areas like the Loess Plateau of Gansu, it is better to apply 200 m³/ha of irrigation water to get the maximum WSE. While the yield is lower than if 400 and 500 m³/ha is applied, with the lower irrigation amount, the yield increase per unit water application is the highest and the overall yield with the available water is the largest. For example, if 60 m³ of water is stored in the tank, then based on an irrigation schedule of 200 m³/ha, the tank water can be used to irrigate 0.3 ha, resulting in a total yield increase of 474.2 × 0.3 = 142.3 kg, while in the case of irrigating

Table 6.11 Spring wheat yield by irrigation amount in the growing season

Irrigation water amount (m³/ha)	Yield (kg/ha)	Yield increase (kg/ha)	WSE (kg/m³)	Water consumption (m³/ha)	WUE (kg/m³)
0	1,443.4			2,826	0.51
200	1,917.6	474.2	2.37	2,980	0.64
400	2,062.7	619.2	1.55	3,654	0.56
500	2,102.7	659.2	1.32	3,873	0.54

Source: Courtesy of GRIWAC et al. (2002)

Table 6.12 Spring wheat yield under different irrigation schedules

Test treatment	Yield (kg/ha)	Yield increase (kg/ha)	Soil moisture (%) Seeding	Soil moisture (%) Harvest	Consumed water (m³/ha)	WSE (kg/m³)	WUE (kg/m³)
WCK	1,334			12.5	3,410		0.39
WJ450	1,803	469		13.7	3,577	1.05	0.51
WJ900	2,018	684		13.6	3,912	0.77	0.53
WB450	1,934	600	14.3	15.7	3,111	1.34	0.62
WB900	2,096	762		15.9	3,517	0.96	0.6
WJB450	2,270	936		12.6	3,839	2.09	0.59
WJB900	2,375	1,041		13.8	4,010	1.16	0.59

Note: WCK is Non-irrigated wheat, WJ450 Wheat irrigated in jointing stage with water of 450 m³/ha, WB450 Wheat irrigated in booting stage with water of 450 m³/ha, WJ900 Wheat irrigated in jointing stage with water of 900 m³/ha, WB900 Wheat irrigated in booting stage with water amount of 900 m³/ha, WJB450 Wheat irrigated in jointing and booting stages with total water amount of 450 m³/ha, WJB900 Wheat irrigated in jointing and booting stages with total water amount of 900 m³/ha.
Source: Courtesy of GRIWAC et al. (2002)

with 400 m³/ha and 500 m³/ha, the land area that can be irrigated from the tank water would be only 0.15 and 0.12 ha, respectively. This results in a lower total yield increase of 92.9 kg using an irrigation rate of 400 m³/ha and 79.1 kg for 500 m³/ha. Since more land can be cultivated the benefit of using a lower irrigation rate of 200 m³/ha is clear, with total yield exceeding those obtained using irrigation rates of 400 m³/ha and 500 m³/ha by 49.4 kg and 63.2 kg, respectively.

It is also important to determine the optimum timing and frequency for irrigation. The results of investigations to determine this are shown in Table 6.12.

From Table 6.12, we can derive some useful results:
- The water applied at the booting stage produces a higher yield and WSE than that at the jointing stage. The former yield is higher than the latter by approximately 7 per cent and 4 per cent for irrigation rates of

450 m³/ha and 900 m³/ha, respectively. WSE and WUE at the booting stage are higher than at the jointing stage by 28 per cent and 25 per cent and 22 per cent and 13 per cent, respectively.
- An irrigation rate of 450 m³/ha results in lower yield but higher WSE than the irrigation rate of 900 m³/ha. To get greater total benefit, the lower irrigation rate is preferable.
- In both cases for irrigation rates of 450 m³/ha and 900 m³/ha, providing two applications, one at the jointing stage and one at the booting stage, results in a higher yield and WSE than just one application (see Table 6.13).
- Therefore, when the available water amount is limited, then irrigation with 450 m³/ha and separated in the jointing and booting stages provides the best irrigation results. The WSE can be as high as 2.1 kg/m³ and the yield and WUE are higher than rain-fed-only cultivation for spring wheat by 70 per cent and 51 per cent, respectively.

Spring corn

Corn is one of the main grain crops in Gansu. It has a higher yield than spring wheat because of the greater available rainfall during its longer growing season. Seeding of corn takes place in mid-April and harvest takes place in late August to early September, the growing season lasting for 140 days. The crop's water demand in the whole growing season is about 460 mm. Natural rainfall in this period is around 320 mm so the water deficit amounts to 140 mm, occurring mainly in late May and June.

Farming practice shows that corn needs less water at the immature stage of the young plant. It has higher tolerance to water stress in this stage and a mild water deficit even benefits the deepening of the root system, thus enhancing its future resistance to drought. The most critical period for water demand comes in the flowering stage when crop growth is very sensitive to water shortage. In this period the WSE can be as high as 3.9 kg/m³. The effect of different irrigation schedules on crop yield is shown in Table 6.14.

Table 6.13 Percentage yield and WSE increases for two applications compared to one application at either jointing or booting

Irrigation amount (m³/ha)	% yield increase compared to		% WSE increase compared to	
	One application at jointing	One application at booting	One application at jointing	One application at booting
450	26	17	99	56
900	18	13	51	21

Table 6.14 Corn yield under different irrigation schedules

Test treatment	Yield (kg/ha)	Yield increase (kg/ha)	Soil moisture (%) Seeding	Soil moisture (%) Harvest	Consumed water (m³/ha)	WSE (kg/m³)	WUE (kg/m³)
CCK	7,131			12.3	4,012		1.8
CJ600	9,270	2,139		12.3	4,605	3.6	2
CJ1200	8,342	1,211		13.2	4,997	1.1	1.7
CB600	9,440	2,309	14.3	11.4	4,820	3.9	2
CB1200	10,000	2,869		12.7	5,109	2.4	2
CJB600	9,066	1,935		12.3	4,603	3.2	2
CJB1200	9,507	2,376		13	5,051	2	1.9

Note: CCK is Corn without irrigation, CJ600 Corn irrigated in jointing stage with water of 600 m³/ha, CJ1200 Corn irrigated in jointing stage with water of 1,200 m³/ha, CB600 Corn irrigated in big bell-mouthed stage with water amount of 600 m³/ha, CB1200 Corn irrigated in big bell-mouthed stage with water amount of 1,200 m³/ha, CJB600 Corn irrigated in jointing and big bell-mouthed stages with total water of 600 m³/ha, CJB1200 Corn irrigated in jointing and big bell-mouthed stages with total water of 1,200 m³/ha.
Source: Courtesy of GRIWAC et al. (2002)

From Table 6.14, we can also draw some useful lessons on irrigation scheduling for corn:

- When the irrigation is applied only once, then irrigation in the big bell-mouthed stage is better than in the jointing stage. WSE of the former is higher than that of the latter for irrigation amounts of either 600 or 1200 cubic metres per hectare. The WUE for irrigation water of 1,200 m³/ha when irrigating in the big bell-mouthed stage is higher than that in the jointing stage by 13.8 per cent. However, the result is the opposite for irrigation water of 600 m³/ha. Irrigation in the big bell-mouthed stage gives a slightly lower WUE, maybe due to some random errors in testing.
- For an irrigation quota of 1,200 m³/ha, applying water twice (in the jointing and big bell-mouthed stages) gives a higher yield by 11.3 per cent than with the same irrigation quota but concentrated only in the jointing stage. However, separated irrigation produces a lower yield than with the same water amount but concentrated in the big bell-mouthed stage. When the irrigation quota is only 600 m³/ha, concentrating the irrigation water in one stage (either jointing or big bell-mouthed stage) gives a better result. This further verifies the big bell-mouthed stage as the most critical for irrigation.
- The yield at irrigation rates of 1,200 m³/ha is higher than at 600 m³/ha by 5–11 per cent, except in the case of irrigation in the jointing stage. However, the WSE for the smaller irrigation quota is higher than that of the larger quota. So it is preferable to use less intensive irrigation but have more irrigated land in production when using an RWH system.

- Irrigation of 600 m³/ha concentrated in the big bell-mouthed stage provides the best irrigation schedule. The WSE is as high as 3.9 kg/m³. The yield and WUE are higher than for the rain-fed-only spring corn by 2,309 kg/m³ (30.4 per cent) and 0.2 kg/m³ (11.1 per cent), respectively.

Millet

A test on different irrigation timings was conducted on millet crops. The results are shown in Table 6.15.

From Table 6.15, it can be seen the best period for irrigation is at the heading stage as this marginally produces the highest WUE.

An investigation into the effect of irrigation quota on yield was also conducted. The results of investigation are shown in Table 6.16.

From the test, the marginally best irrigation quota is 300 m³/ha.

It should be stressed that the above examples are only strictly relevant to the Loess Plateau of Gansu and to other areas with similar climate and soil conditions. These case studies are not intended to provide a cure-all irrigation schedule for other areas and these results cannot be simply transferred and applied. However, the examples do highlight some important principles and underscore the importance of formulating precise irrigation scheduling, especially when using finite water sources such as stored rainwater as the irrigation source. The results prove that wise irrigation scheduling can enhance the efficiency of rainwater use. The key lesson here is that we should not only focus our efforts on building the RWH systems, but also pay careful attention to how the stored water is efficiently managed to ensure the maximum benefit is achieved.

Table 6.15 Millet yield by irrigation timing

Irrigation period	Yield (kg/ha)	WUE (kg/m³)
Jointing	4,010	1.5
Heading	4,257	1.55
Flowering	4,107	1.54
Non-irrigated	3,755	1.51

Source: Courtesy of GRIWAC et al. (2002)

Table 6.16 Yield of millet by irrigation quota

Irrigation quota (m³/ha)	Yield (kg/ha)	WUE (kg/m³)	WSE (kg/m³)
0	3,755.8	1.52	
100	3,862.1	1.53	1.06
200	3,958.3	1.53	1.01
300	4,114.6	1.54	1.2
400	4,117.6	1.51	0.9

Source: Courtesy of GRIWAC et al. (2002)

PART III
Rainwater harvesting and environmental management

CHAPTER 7
Small watershed management

Background

China is a country facing some of the most serious soil erosion in the world, affecting over 37 per cent of its land area. Every year around 5 billion tonnes of soil are eroded, of which 2 billion tonnes enter the sea. Among all the areas affected by soil erosion, the Loess Plateau and mountainous areas in northwest China are the most seriously impacted. As previously outlined, the Loess Plateau of Gansu Province is characterized by water scarcity, low agricultural productivity, and a history of impoverishment of the rural population. Another feature of the area is intense soil erosion, totalling 6,000–15,000 t/km² annually. There are several factors affecting soil erosion in the loess area of Gansu:

- Most of the land surface is bare or only sparsely covered with vegetation and 70 per cent of the total land area slopes between 5 and 50°, averaging around 10–15°.
- In the rainy season, most of the rainfall is in the form of storm events that cause intense soil erosion.
- The widely distributed loess soils are mainly composed of silt and fine sand that has low cohesive strength. Clay content is usually less than 30 per cent. The density of loess soil mostly ranges between 1.1 and 1.3 t/m³ i.e. low, and the loose structure makes it easily eroded.
- The loess soil has a special property of wet subsidence; this means it will subside when wetted. The soil also has a high infiltration rate due to the presence of vertical fissures. When heavy rainfall occurs, water flows through these fissures causing the loess soil to subside or even collapse.
- Human activities also exacerbate soil erosion. Due to low soil productivity, farmers use every piece of land for cropping, even those on steep slopes. Overgrazing and uprooting of straw and brush for household fuel further diminish the vegetation cover, causing more soil erosion and land degradation. There is a Chinese saying: 'the poorer they were, the more they reclaimed, the more they reclaimed, the poorer they would be'.

Soil erosion causes a number of adverse impacts on agricultural production, local eco-systems, and the wider environment.

One of the impacts of soil erosion on agriculture production stems from the gullies and ravines caused by erosion, triggering land slips and collapses, making the land unusable. Stripping off the surface soil causes loss of valuable nutrients. The annual nutrient loss per hectare in 1993 was estimated at 24 kg of total nitrogen, 223 kg of potash, and 49 kg of phosphate. Organic matter decreases by 35–67 per cent after erosion. Soil erosion also damages the soil's granular structure, thus lowering its moisture retention and ventilation capacity.

On environmental impacts, soil eroded from the watershed causes sedimentation in the main tributary of the Yellow River (Huang He), with about 1.6 billion tonnes entering the river annually, mainly from the Loess Plateau area. This causes 8–10 cm silting of the riverbed in the lower river each year. In its lower reaches the river flows several metres higher than the adjacent cities and farmland, creating a threat to people during flooding.

The nutrient loss from the soil is a source of pollution of rivers and lakes downstream, causing eutrophication and intensifying water shortages. The increasing application of chemical fertilizers and biocides for agriculture in recent years has further exacerbated pollution.

The adverse impacts of soil erosion have reduced agricultural output, perpetuated poverty, and exacerbated environmental degradation in many parts of China. On the Loess Plateau of Gansu, soil erosion has become the main factor restricting social and economic development and impeding conservation efforts. As a result serious attention is being given to erosion control measures both in Gansu and throughout China. This is recognized as a strategic intervention in addressing poverty alleviation, sustainable economic development, and environmental conservation in the 21st century.

Past experience has shown that when carrying out erosion control, working at the small watershed level produces the best results. Here, small watershed means a river basin with an area of less than 100 km². An integrated management approach towards these watersheds has proved to be the most effective way of reducing soil erosion and land degradation. Urgent intervention is also essential, as the Loess Plateau and mountainous areas in Gansu are facing a serious challenge from soil erosion. More than 95 per cent of the loess areas in Gansu suffer from soil erosion. In recent decades, integrated measures for controlling soil erosion have been taken in more than 600 small watersheds with a total area of 12,000 km². The two main activities in watershed management include: terracing and planting trees and grasses for erosion control on slopes, and building check dams[12] and silt arrestors. In the past 20 years, RWH has also become an integral part of watershed management.

Scientific approach to soil erosion control

Past experience with soil and water conservation in China has shown that for cost-effective results, an integrated and scientific approach should be adopted, tailored to the local topographic and climate conditions.

Topography

Two basic types of topography characterize the watersheds in the loess areas of Gansu Province. The first type of watershed comprises a gullied loess plateau. The area of a plateau can range from several hectares up to hundreds of square kilometres with the surfaces dissected by gullies. The second watershed type is gullied loess hills. The elevation difference from the top to the outlet of these hill watersheds is usually 200–400 m and the mean gradient ranges between 12 and 17°. The density of the gullies and ravines ranges from 1 km to nearly 4 km per square kilometre.

In plateau watersheds, the main cause of erosion is runoff from the plateau itself during heavy rainfall, resulting in gully erosion.[13] Over thousands of years this has formed numerous gullies and ravines. Through continuous scouring action these gullies have become wider and deeper and cut back, dissecting the plateau into small areas. As a result, the slopes have become unstable, leading to numerous landslips.

To minimize runoff, the gentle slopes of the plateau are terraced to retain most of the rain. The terraces comprise contour strips 30–50 m in width, with trees planted along the edges to retain more runoff. The runoff often concentrates along roads that eventually become gullies. Both sides of the road are therefore levelled into terraced plots onto which runoff from the road is diverted and trees are planted. On hillsides, with gradients less than 25°, the slopes can be terraced for cropping or planting fruit and timber trees. On steeper slopes, the land has to be levelled into strips (contour planting or contour strips) or plots, for example the fish-scale plot or levelled plot on which trees or bushes can be planted. In the valley bottoms the banks are often steep. Trees and/or bushes are planted in the plots or contour strips. In the bottom of gullies, willow, earth, or stone check dams are usually built to retain silt. In the main gully, dams are built for arresting silt and after operating for some years, flat land is formed behind the dams.

For the second type of watershed, there are no flat plateau areas, only small hill tops. Measures for erosion control again involve planting trees and/or bushes in contour strips or on levelled plots. The contour strips stop or significantly reduce erosion as well as retaining rainfall that helps trees and crops to survive, so this is also a form of RWH. As with the plateau watersheds, terracing and planting of trees, bushes, and grasses are undertaken on the hillsides and check dams are built in the gullies for retaining sediment and forming flat land. Figure 7.1 shows the arrangement of erosion control measures in the Qianjiagou watershed, which is representative of gullied plateau areas. Figure 7.2 shows erosion control measures in the Baozigou watershed, representative of the areas of loess hills with gullies.

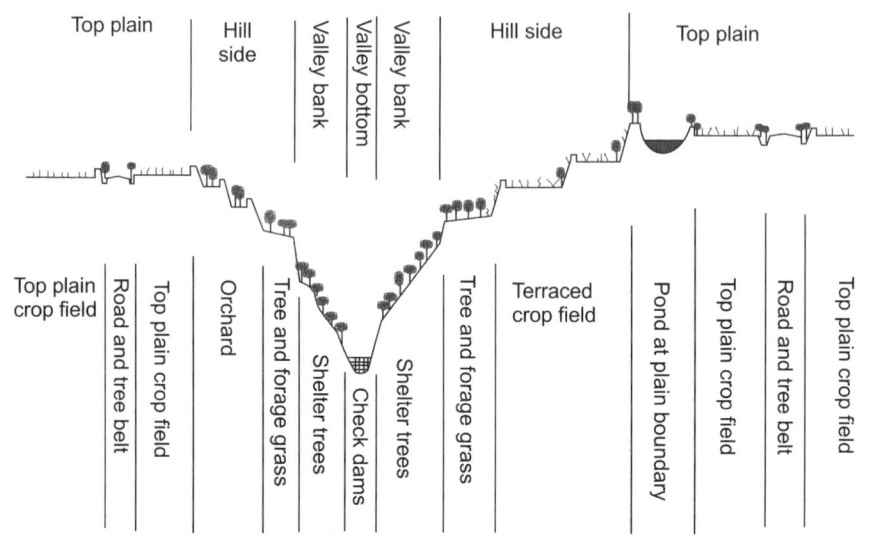

Figure 7.1 An arrangement of erosion control measures in Qianjiagou watershed (representing areas of loess plateau with gullies)

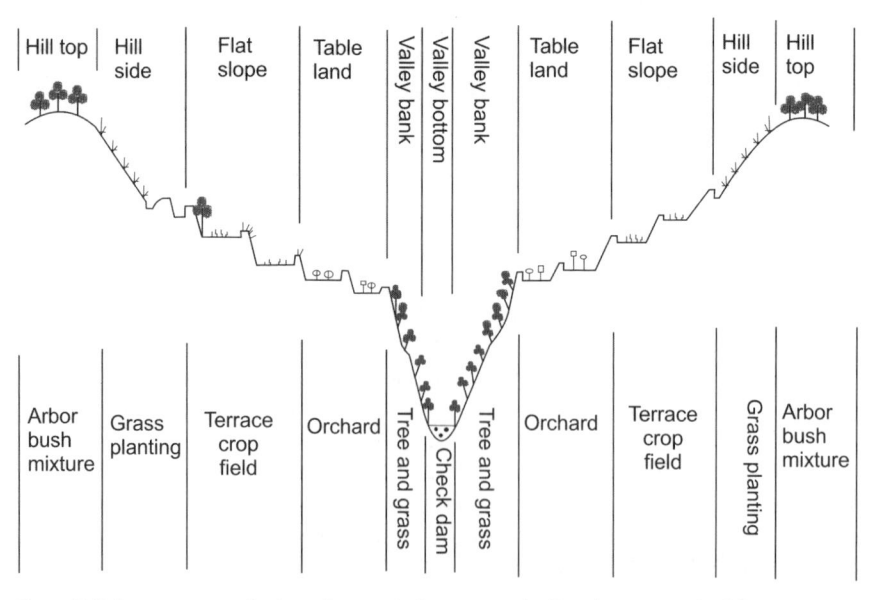

Figure 7.2 An arrangement of erosion control measures in Baozigou watershed (representing areas of loess hills with gullies)

> **Box 7.1 3-W model in the Jiuhuagou watershed**
>
> Jiuhuagou watershed is at the centre of the loess hilly area. It has an area of 83 km². The annual precipitation is 380 mm and the potential evaporation is 1,500 mm. Gullies and ravines criss-cross the watershed with a density of 2.7 km/km². This is one of the areas with the most serious soil erosion in the province and has an annual erosion rate of 5,400 t/km². This results in a nutrient loss per hectare of 519 kg of organic matter, 39 kg of total nitrogen, 34 kg of total phosphorus, and over 7 kg of quick effective potassium. As a consequence, agricultural production has been low and the local people have remained poor. Until 1996, the average crop yield in the area was only 600 kg/ha. The land was mostly planted with grain crops and due to lack of water and poor soil very few cash crops were planted. The annual income per capita was only $110. In the 1960s several experts visited the area and afterward they affirmed that this area was not suitable for human settlement.
>
> Harnessing the potential of the watershed started in the 1970s and, in 1987, the watershed was assigned support from the provincial small watershed management programme. The technicians and local people developed what became known as the 3-W model for soil erosion control. These Ws denote to Wear a cap (forest) on the hill top, Wrap a belt (terrace or contour planting) on the hillside, and Wear a boot (dams) at the toe of the slope.
>
> By 2000, around 1,000 ha of sloping land had been reclaimed by terracing. For slopes steeper than 25°, mixed trees and bushes were planted on an area of over 1,440 ha, and alfalfa was planted on nearly 929 ha of eroded hillsides. The building of 1,500 S*hiujiao* rainwater tanks was also completed, supplying irrigation to around 330 ha of land.
>
> Water conservation measures for agricultural production were also adopted, including the use of plastic sheeting, micro-catchments, water-saving irrigation, and irrigation during seeding. The cropping pattern was modified to adapt to the natural rainfall conditions by planting more corn and reducing the amount of wheat grown. Monoculture cropping systems were diversified and more cash crops were cultivated.
>
> During the project, the percentage of land on which erosion control measures were adopted increased from 45 per cent to 86 per cent in the five years between 1987 and 1992. Runoff was reduced from 17,000 m³/km² to 1,560 m³/km² and erosion rates decreased from 5,400 t/km² to 915 t/km². The land use proportion of afforestation and animal husbandry increases from 30% and 8% before harnessing to 44% and 26%, respectively. Land utilization increased from 63 per cent to 86 per cent. As a result per capita incomes increased by over 97 per cent and annual per capita grain production increased from 410 to 654 kg.

Climatic conditions

A study of the cost effectiveness of erosion control measures in Dingxi County (annual precipitation round 430 mm) by the Anding Soil and Water Conservation Station in the 1990s shows that climatic factors such as wind and the orientation of the slope are important considerations when selecting erosion mitigation measures. The main conclusions are:

- The wind speed in the watershed of the loess areas in Gansu is relatively high. It was found that when the speed exceeds 2 m/s the wind has negative impacts on plants, and especially agricultural crops. Under windy conditions, plant transpiration is restrained due to lower soil

moisture, which is unfavourable for growing broadleaf trees. However, the impact on shrubs such as the *Caragana korshinskii* and *Tamaricaceae* is less significant.
- The wind speed varies for different locations on the hill slopes. At the upper and middle levels it is much higher. Therefore, in the upper and middle zones of the watershed, shrubs are more suitable than broadleaf trees.
- Owing to the wind and solar radiation, soil moisture on hill tops and south-facing (sunnier) slopes is lower than that on shaded slopes. In areas with annual precipitation of less than 350 mm, planting of broadleaf trees is not suitable.
- Crop yields of the upper part of the watershed would be lower than yields of the middle and lower part.
- The orientation and location of the slope has no significant impact on the yield of grass, which is suitable to plant at any location.

Terracing

Terracing changes the original slope of the land to produce horizontal steps. Terracing in Gansu can be traced back a thousand years. At present, it is still used as an important measure to enhance productivity of rain-fed agriculture by conserving fertile surface soils and increasing soil moisture. It remains one

Photo 7.1 Terraces with plastic sheeting in Zhuanglang County

SMALL WATERSHED MANAGEMENT   101

of the main activities in small watershed management for controlling soil erosion and conserving rainfall. Over the past 50 years 1.9 million ha of the sloping land has been terraced, amounting to 65 per cent of all sloped land in the province. Photo 7.1 gives a view of the terraces in Zhuanglang County, which has been given the title of 'Terraced County' by the state.

Terrace design

Terracing in the loess areas in Gansu can be divided into two types: continuous terracing and terraces interspersed with the original slope, as shown in Figure 7.3.

Continuous terracing is done in areas where rainfall is relatively high and land availability low. It consists of level terrace land and a bank with a low dike at the top. The crop land should be levelled or have an inverted slope of about 1 per cent for increased rainfall retaining capacity. The dike provides a reservoir for storing a certain amount of rainfall. The outer slope of the bank should be flat enough to meet the demand of slope stability (see Figure 7.3).

Terraces in between the original slope are used in areas of low rainfall and where land is not in short supply. The slope in between the terraces is designed to concentrate rainwater runoff and direct it onto the lower terrace where it increases the soil moisture content.

An additional type of terrace, the slope or contour terrace, can be formed by continuous tillage over many years, turning the soil always in the same direction to gradually flatten out gently sloping land.

To design a terracing project, the overall scope of the project has to be decided and a plan, including all the terraces to be constructed, should be formulated. In the loess areas in Gansu, a hillside bordered with gullies deeper than 5 m or with roads can provide the boundary of a terrace project. Usually a terracing project consists of several or tens of pieces of terraced land. If the original slope is gentle, the size of each terrace can be 0.5–1 ha, while on steep slopes, just 0.1–0.2 ha. The project site should not be located in areas where landslips may occur. The gradient of the ground should be less than 25°

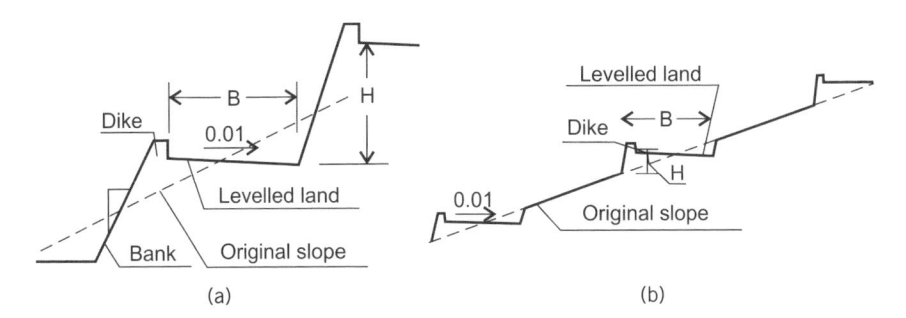

Figure 7.3 Two types of terrace: (a) continuous terracing and (b) terraces with the original slope partly retained

(preferably less than 15°) and the soil depth at least 3 m. Since the terracing work will be carried out between harvest time and the next seeding, the crops in the field to be terraced should be the same or have the same growing period.

The terraced land is the unit of cultivation. Other components of the project include the access road, forest belts, and any canals or irrigation systems that have to be built before any terracing work begins.

The access road, normally 3–4 m wide, links the village and the field for movement of all materials, machinery, tools, products, and people. The maximum road gradient should not exceed 15 per cent and a turning radius of no less than 12 m needs to be incorporated. When the hillside gradient is less than 15 per cent, the road can be aligned perpendicular to the land contour. Otherwise, it needs to zigzag up the slope.

The forest belt is located along the roadside bank, on the bank toe, or along the small dike next to the terrace, depending on the moisture conditions. Bushes and grass may also be planted on the bank and/or dike. However, in areas with annual rainfall of less than 350 mm, trees and grass should not be planted on sunny slopes as these become too dry.

After planning the terracing project, the design of the terraces is carried out. The design parameters include the width B, the height of the bank H, the angle between the bank and the horizontal β, the size of the dike, as well as the length of the terrace.

The height of the terrace H is related to the width B for a given slope. The greater the height, the greater the width, and vice versa. The following factors are taken into consideration when sizing the terrace:

- To reduce the cost of terracing, note that terracing with greater bank heights requires more (labour and/or machine) time than that with smaller heights.
- The smaller the bank height, the greater the number of terraces and the area occupied by banks per unit of productive area, leading to greater percentage land loss. To reduce land loss, higher terraces are preferable. Wider and higher terraces allow cultivation, either manually or using machinery, to be more efficient.
- Terrace stability depends on the shear strength of the soil and the height of the bank. Greater shear strength allows higher terraces to be built.

There can be different width (B) and height (H) combinations for a terrace with certain slope and soil conditions. The Gansu Soil and Water Conservation Institute carried out a study in the 1990s to optimize the size of terraces. A multi-objective model was developed to minimize the cost and maximize the benefit with the following variables:

- cost (in the form of labour days) of terracing versus sizing;
- land loss or benefit by terracing per hectare of terrace versus sizing;
- estimation of cultivation cost of machinery/labour versus sizing.

The bank slope depends on the bank height and the soil shearing strength. Stabilization of the bank is a fixed constraint when solving the multi-objective model to determine terrace parameters.

To estimate the land loss and benefit due to terracing, several factors must be further considered. First, the solar radiation received by the surfaces changes after terracing due to the change of the land slope. After terracing the radiation received increases for slopes oriented to the north and reduces for slopes facing south, east, and west. This phenomenon causes changes in the soil moisture conditions and thus productivity. Second, terraces retain much more moisture and nutrients than sloping land.

In the multi-objective optimization model, all the variables are expressed in cash and a calculation period of 30 years is adopted for maximizing the benefit (minimizing the cost). Using local parameters, the Gansu Soil and Water Conservation Institute achieved the results shown in Table 7.1 on the sizing of terraces.

These results are based on the local geographic, climatic, and economic conditions of the loess areas in Gansu.

Sizing of dikes involves determination of height, the inner and outer slopes, as well as the width of the crest. The height of the dike is determined by the volume of runoff generated from the terrace area by a storm with a 20 year recurrence interval. In Gansu, usually a 20–30 cm high dike is adequate. The bank and dike are built of compacted earth. The width is normally 30–40 cm and the inner slope is 1:1. The outer slope of the dike is the same as the bank. Sometimes the bank is planted with grass, in which case the slope is not compacted but simply hit with a shovel. This is called a 'soft' bank and has a slope of 1:1 (45°). In this case the dike has an outer slope of 1:1 and the inner slope of 1:2 (26.5°).

Terrace construction

Terracing is done to level the slope into horizontal strips of land, as shown in Figure 7.4. The lower part of the slope is filled and the upper part cut away.

A typical terrace project usually involves several to tens of terraces as the basic unit. The number of terraces in the project should be designated. To do this, a baseline (perpendicular to the contours) is identified marking the area

Table 7.1 Results for optimized terrace parameters

Orientation	South					North					East and west				
Original Slope degree	5	10	15	20	25	5	10	15	20	25	5	10	15	20	25
H (m)	2.5	3.0	3.2	3.3	3.4	2	2.4	2.5	2.6	2.7	2.5	3	3.2	3.3	3.3
β (degrees)	82	78	76	75	75	86	83	82	81	81	82	78	77	76	71
B (m)	28.6	17.1	12	9.1	7.2	22.3	13.4	9.4	7.2	5.7	28.1	16.8	11.8	9	7.1

to be terraced and the distance between the top and bottom of the baseline measured. Using the gradient and the orientation of the hill slope, the optimal height H and width B of the terrace can be found using Table 7.1. The number of the terraces can be determined with the following equation:

$$N = \frac{H_{Total}}{H_t} \quad (7.1)$$

where N is the number of the terraces, H_{Total} is the difference in elevation between the two ends of the baseline, and H_t is the optimum height of the terrace. The length of the slope can then be divided into equal parts by the number of terraces, and the boundary of every terrace can be marked out on the ground. Each terrace is divided into grids and measurements are taken to find out the elevation of each node. The depth of cut and fill of each node is then calculated.

During terracing, it is very important to conserve the fertile top 30 cm of soil. Otherwise, the fertility of land will not recover for several years and yields will be lower than from the original field. Conserving the top soil during terracing requires about 150–200 additional labour days per hectare and Gansu farmers have developed several ways to do this.

- Method 1: First all the top soil is moved to the middle part of the slope where little fill and cut is needed. After levelling the slope by cutting and filling (Figure 7.4), the top soil is then spread back over the new terrace. Since moving top soil from the lower part of the slope upward to the middle is very hard work and labour intensive, this method is best done by machine.
- Method 2: The slope is divided into four to five lateral sections. Top soil from the first section is not conserved. After levelling the first section, the top soil of the second section is moved to cover the first and so on. The top soil from the whole terrace is recovered section by section, except for the first one. This method requires less labour.
- Method 3: The lowest terrace in the terrace group is first levelled without recovering the top soil. Then the top soil from the slope above is moved to cover the finished terrace before the next terrace is constructed, and the process repeated. This way whole project will have its top soil recovered with less labour/machinery, except for the highest terrace.

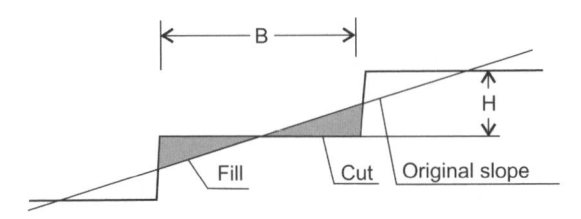

Figure 7.4 Section of cut and fill in terracing

- Method 4: The slope is divided into lateral sections with widths of 4–5 m. For every two sections, the top soil is excavated and moved from one section and piled on the neighbouring section. Land levelling is done in the first section where the top soil has been removed. After that, the top soil piled on the second section is returned to the first section that has now been levelled. Then the second section is levelled with its top soil piled temporarily on to the first section. Finally the top soil is replaced on the second section.

The bank of the terrace should be compacted, usually manually, using wooden forms made of timber. In this case, the slope of the bank can be around 70–85° (see Table 7.1). Sometimes the bank is made just by piling the soil from the cut sections and compacting this with a shovel. In this case the slope should be around 45°.

Until quite recently most terracing in Gansu was done manually. Typically, terracing of loess slopes required about 750 labour days per hectare and the government has provided subsidies to farmers to encourage this. Since the terraced land is used by the farmers, who are the direct beneficiaries, the subsidy has been very low. Although the subsidy has increased with time, it remains low. In recent years, terracing has increasingly been done by machine. Since labour costs have increased in China, terracing using machines has now become more economical. For example, at a terracing project in the Xipinggou watershed in Guanghe County implemented in 1991, 30 ha of terracing took 300 machine days using a bulldozer plus 3,000 labour days from workers. The total cost for each hectare is estimated at about $1,400 at 2008 prices. The use of machinery for terracing operations requires a terraced field width of no less than 7 m. Therefore mechanized terracing is suitable only on slopes with gradients of less than 20°. Use of machines allows for recovery of top soil using Method 1.

Terrace operation

The benefits of terracing include both ecological and environmental benefits, as well as social and economic benefits. Through terracing, soil erosion can be controlled, productivity enhanced, and people's lives improved. When sloped land becomes horizontal, soil and water loss is reduced. In the next section we discuss the further control of soil erosion by planting trees and grass. Improving the productivity of the land is sometimes harder than expected. A key point is to ensure recovery of the top soil to maintain fertility.

In Gansu a target of 70 per cent of top soil recovery is used to ensure a significant increase in crop yield. However, conservation of the top soil needs a lot of labour/machinery and careful construction procedures. Sometimes, the proportion of the top soil recovered is less than 70 per cent, the top soil becoming mixed with the sub-soil during terracing. In this case, for a period of years the yield on the terrace may be lower than in the original field. To ensure quick increase of yields, specific measures must be taken.

First, fertilizers including manure and/or chemical fertilizers such as nitrogen, phosphorus, or potassium must be applied to the fields. Applying animal manure increases the organic matter content as well as the essential nutrients to the soil. Crop resistance to drought is also closely related to the organic content of the soil. In addition, applying animal manure, planting green manure crops after harvest time, and leaving stubble in the fields are all effective measures to increase the soil's organic content.

Applying chemical fertilizers can rapidly increase the nutrients in the soil of newly built terraces. A demonstration on thousands of hectares of land in the early 1990s showed that for the application of every 1 kg of nitrogen, wheat crop yields per mu increased by 19 kg, and for each 1 kg of phosphorus i.e. P_2O_5 by 13 kg. Testing showed that applying fertilizer in three periods of base, seed, and top applications is more effective than a single application. The demonstration also revealed that to keep a balance between nitrogen and phosphorus, use of animal manure is important. It was found that if the top soil was not recovered after terracing, the organic content and the total nitrogen of the soil in the fill part were 10 per cent higher than in the cut part, from which top soil was depleted. In this situation, to even out fertility in the terraces, fertilizer application in the cut section should be double that in the fill part.

A second measure is deep ploughing, which can lower the soil density by up to 11 per cent and increase the porosity by around 6 per cent. Deep ploughing is undertaken in the summer and autumn, which are the hot and rainy seasons. More voids in the soil at this time allow for greater rainfall infiltration and moisture storage. It also benefits the activity of micro-organisms and enhances the activity of soil enzymes. Deep ploughing creates a granular soil structure. However, too frequent and too deep ploughing can lower the fertility of the top soil, causing more water loss. Practical experience has shown that the best depth to plough is 30–33 cm in semi-arid areas and 20–25 cm in sub-humid areas, and up to three times after harvest.

A third measure is three-yearly crop rotation. The first crop planted after terracing in semi-arid areas should be a drought-resistant variety. The usual rotation is potato, pea, and hyacinth bean. In sub-humid areas where moisture conditions are better, the key concern is soil fertility. Here beans, potato, and cereals are usually rotated. If deep ploughing and significant fertilizer (manure and phosphorus) application are undertaken after terracing in the summer, then winter wheat may also be cultivated.

In Dingxi Municipality, three kinds of crop rotation have been tested, namely: (1) pea–wheat–linseed; (2) hyacinth bean–wheat–linseed; and (3) potato–wheat–linseed. Over three years, the yield of the third option was found to exceed the other two by 16 per cent and 60 per cent, respectively.

The benefits of terracing

Terracing plays an important role in watershed management. The benefits of terracing include conservation of the environment through reducing soil erosion and land degradation, and improving agricultural productivity and thus increasing people's living standards. In addition, terracing has benefits for both the downstream water basin environment and population.

Benefits for the environment. Terracing has significantly improved soil and water conservation. The terraces in Gansu are designed to retain runoff from storms as large as 1-in-20-year events. By 1993, the terraced land in Gansu amounted to over 1 million ha. A study in 1994 estimated that the terraces in Gansu could retain around 770 million m^3 of rainwater and 108 million tonnes of silt annually. The results showed on average about 720 m^3/ha (72 mm) of water was being retained annually by the terraces, producing a reduction of approximately 100 tonnes per hectare of soil erosion. The report estimated that the runoff on slopes in the loess area of Gansu ranges between 600 and 1,050 m^3/ha annually, so most of the runoff can be retained in the terraced fields.

Baozigou watershed located at the Zhuanglang County in Gansu has an area of 18 km^2. A study of soil and water conservation was conducted in this watershed from 1988 to 1992. The study monitored rainfall, runoff, and soil erosion for the different landscape types. The results showed that the terraces, together with trees, bushes, and grass planted on the banks and dikes, plus pits for planting trees, retained most of the rainfall runoff and stopped 100 per cent of soil erosion.

Benefits for agriculture productivity. In the loess areas of Gansu the annual rainfall is only 250–550 mm. The unfavourable seasonal distribution of precipitation further aggravates water shortages, which are the main constraint for agricultural production. Terracing can mitigate the problem to a considerable extent. Investigations in Jingchuan County, monitoring moisture in the soil profile from the surface to a depth of 180 cm in wheat fields, showed that on average terraced land can provide 58 mm more soil moisture annually than the original sloping land. Testing also shows that the evapotranspiration from wheat on terraces in the growing period is 54 mm greater. After the growing season, the soil moisture in the terrace is lower and evaporation from the bare fields on the terraces is almost 7 mm less than that on sloping land.

After terracing, nutrient loss due to soil erosion is greatly reduced or even stopped completely. Fertility lost due to the removal of top soil can be recovered by careful fertilizer application, deep tillage, and crop rotation as described above. Farming operations and especially those involving machinery are much easier, safer, and more efficient on terraced land, which also creates the opportunity for irrigation using RWH systems. The experience in Gansu shows that terracing can increase crop yields by 40–50 per cent compared to unterraced slopes.

Investigations on 416 ha of terraced land and 91 ha of sloped land in Jingchuan County, found that the crop yields over eight years from 1987 to 1994 were on average 54 per cent higher on the terraced land and the increase was higher in the drier than in the wetter years. From 1988 to 1990, the yearly rainfall total ranged between 564 and 680 mm, greater than the mean annual rainfall on the slopes by between 12 and 129 mm. The yield on terraces was 42 per cent higher than on the slopes.

In dry and very dry years, while overall crop yields will be down, the difference between terraced and non-terrace land is even greater. From 1991 to 1994, the annual rainfall ranged between 1.5 and 36 per cent less than the average, but the yields on terraced land were 56 per cent higher. In 1995, a 1-in-60-year drought occurred and the annual rainfall was only 46 per cent of the norm. The crop yields on both the terraced and sloped land dropped significantly. Nevertheless, the terraces yielded almost 85 per cent more than the sloped land.

In addition to yield increase, the terrace banks and dikes provide the space and opportunity for farmers to plant trees, grasses, and also cash crops. The area of the banks and dikes amounts to between 11 and 19 per cent of the terraced area. In addition to the economic benefit, the plants on the bank and dike help to further reduce soil erosion and increase the infiltration rate of the bank soils, as well as their water-retaining capability.

Growing trees and grass

Technical experiences of tree and grass planting in the loess area of Gansu

Tree and grass planting are also key activities undertaken in watersheds to reduce soil erosion, but to be effective there are some essential points to be considered. These include land preparation for rainfall concentration, the proper arrangement of trees and grasses, and planting trees adapted to local conditions.

Land preparation. Since the rainfall in the loess area of Gansu ranges between 300 and 550 mm, which is not enough for tree growing, rainwater concentration measures are essential to ensure the survival and fast growth of trees. Land preparation prior to tree and grass planting has two functions. First, to concentrate the rainfall to ensure the survival and healthy growth of the plant. Second, by retaining more runoff, the levelled land and ditch or pit helps to reduce soil erosion. Land preparation techniques include construction of terraces, contour strips, contour ditches, fish-scale pits, and deep pits. The particular measure chosen depends on local climatic conditions and topography.

Terracing is suitable on slopes of 25° or less. The width of the terraces ranges between 6 and 12 m and the length can be up to 100 m. Since terracing is expensive in terms of labour and/or machinery, trees planted on the terraces are usually high-value fruit or ornamental trees. Contour strips are used for planting trees when the slope ranges between 15 and 25°. The width of the strips are usually 1–1.5 m, with an inverse slope of 3–5°.

Levelled ditches are also used for slopes ranging between 15 and 25°. The top and bottom width of the ditch is 0.6–1 m and 0.3–0.5 m, respectively. Soil cut from the ditch is used for the outside dike. Trees are planted on the inner toe of the dike.

Fish-scale pits are used on steep slopes, often on the bank of the valley or gully. These are semi-circular in shape with a length of 0.8–1.5 m, a width of 0.5–0.8 m, and a depth of 0.3–0.5 m. They are placed along the contour line in a triangular pattern with intervals between rows and pits of 1.5 m and 2 m, respectively.

Micro-catchments (pits). According to the national code, a levelled deep pit should have a width of 0.8–1 m and depth of 0.8–1 m. In current practice, however, it has become much larger, being square or rectangular in shape, 2–3 m across and 1–1.5 m deep. It is a common technique when planting trees in Gansu and is very effective for rainwater concentration.

Proper arrangement of tree and grass planting under different topographic conditions. On hill tops, if the hill is wide and relatively flat, forage crops such as alfalfa can be planted without land levelling. If the hill top is narrow and the slope is steep, then the land has to be levelled into contour strips. Commonly used shrubs include the hardy locust, pea shrub (*Caragana korshinskii*), and shrubby false indigo.

On the plateau top plain, field sizes are typically around 6 ha. Shelter belt trees including poplar or paulownia are planted along the boundaries. To improve moisture availability, trees are planted in pits (micro-catchments) 1.5 m deep to concentrate rainwater runoff. Trees are also planted in pits dug on both sides of the road, with a single row of paulownia at intervals of 2–3 m or two rows of poplar at an interval of 1–2 m. At the origin of a gully where erosion can rapidly develop headwards, a shelter belt with two to three rows is planted and a dike with a depth of 1–2 m is built inside the tree belt to retain the runoff. Fast-growing trees are also planted at the toe of the steep slope below the head of the gully to prevent further headward erosion.

In terraced fields, paulownia is planted in a single row 3 m apart at the toe of the bank. On the dike, shrubs such as shrubby false indigo and *Caragana korshinskii* are planted. On hillsides with slopes steeper than 25°, trees are planted in contour strips with widths of 1.5–3 m. On south-facing slopes and those with good soil and moisture conditions, commercial tree species can be planted. On shaded slopes and with poor conditions, locust or poplar are planted at an interval of 1–2 m. If the slope is steeper than 35°, then the trees are planted in levelled plots. Photo 7.2 shows contour strips on hillsides and the levelled plots for growing trees can be seen in Photo 6.8 (left).

Cultivated steeply sloping land, remote from villages or located at the edge of gullies, usually has poor fertility. This is generally converted from crop cultivation to planting timber trees and/or forage grass. Usually fast-growing trees such as paulownia or poplar are selected, and narrow terracing, contour planting, or micro-catchments (pits) are constructed, depending on the slope. The forage grasses most commonly used are alfalfa and prairie milk vetch.

On slopes with gradients of less than 25°, terraces are built and commercial fruit trees (apricot and pear) or ornamental trees (pine and oriental arbor vitae) are usually planted. Below the hillside, there is often table land or an 'old terrace' with a slight slope and here, close to the villages, orchards are generally established.

The lower slopes of the valley and gully sides are often very steep. Levelled strips can be built on these slopes when they are less than 35°. For steeper slopes, the levelled plot or fish-scale pit is built for planting trees. A mixture of trees (locust, elm, ailanthus, etc.) and shrubs (*Caragana korshinskii*, shrubby false indigo, sea buckthorn, etc.) are often planted on the upper slope where the land is usually poor. A variety of fast-growing trees such as poplar, locust, and paulownia are planted on the middle and lower part of the slope. In the wide valley bottoms, contour strips or levelled plots are built and poplar, willow, and locust trees planted. If the gully bottom is narrow, then check dams are built to stop erosion.

Planting trees adapted to the local conditions. Proper selection of trees is very important in the loess area of Gansu where the annual rainfall is less than

Photo 7.2 Contour planting for trees

> **Box 7.2 Tree and grass planting in the Qianjiagou watershed**
>
> Qianjiagou is a watershed in the Jingchuan County with area of 60.6 km^2. It is part of the Loess Plateau and is composed of the top plain (31.6 per cent), slope under the plateau (28.1 per cent), and the gullies (40.3 per cent). The elevation varies from 1,358 m above sea level at the plateau down to 956 m at the outlet of the watershed. The annual precipitation is 588 mm, of which 55 per cent occurs from July to September, mostly in storms. The soil has a loose structure that, in combination with the relief, heavy storms, and widely distributed loess soil, results in serious soil erosion. The comprehensive reclamation of the watershed started in 1973. At that time the vegetation was sparse, with forest and grassland covering 9 per cent of the total area and annual erosion averaging 8,000 t/km^2. In the 15 years after 1973, a total of 1,650 ha of forest were planted and 866 ha of terraces constructed. By 1987, the total area undergoing reclamation reached 82 per cent. The approach to tree planting was to select a tree variety appropriate to the local conditions, prepare the land properly in advance, and plant the trees at the right time.
>
> In the vicinity of houses, along roads and field boundaries, and where fertility and moisture conditions are good, fast-growing paulownia and poplar trees are common. Trees were also frequently inter-planted with crops. On sunny slopes walnut, persimmon, Chinese prickly ash, or apple trees were usually selected. On steep slopes and in gullies, the locust, poplar, and ailanthus trees that have well-developed root systems and are drought resistant have been used as shelter trees. For the orchards located on north-facing slopes, hawthorn and pear trees are best suited.
>
> The most appropriate time for planting trees depends on the location. Since rain is rare in the spring on shady north-facing slopes, land preparation is done in the autumn for trees planted the following spring. For the sunny south-facing slopes, land preparation is done in the spring and the trees planted in autumn.
>
> Land preparation varies according to the topography. On slopes of less than 15° and on the table land, narrow terracing and contour planting have been adopted. For the steeper slopes with gradients of 15–35°, contour planting has been used, while on the gully sides with gradients of more than 35°, contour ditches and fish-scale pits were adopted.
>
> With these guidelines carefully observed, afforestation has been very successful. In recent years, the tree growth rate has been 21 per cent annually. In the 15 years between 1972 and 1987, the percentage of forest cover increased from 9 to 26 per cent.

400 mm and many trees cannot grow well. While shrubs are often used, drought-resistant trees can also be selected for land reclamation. *The China Regulation of Techniques for Comprehensive Control of Soil Erosion* (China Bureau for Technical Supervision, 1996) for afforestation identifies seven different climatic zones, and different varieties and plant densities for trees and shrubs are suggested for each zone. Different varieties of grasses are also recommended for different climatic and ecological environments. In the loess area of Gansu, the main trees are paulownia, poplar, locust, Chinese pine, Chinese arbor vitae, and elm, as well as pear, hawthorn and Chinese prickly ash for fruit. The main grasses are alfalfa and clover, and prairie milk vetch (*Astraglus adsurgens*).

Scientific tree and grass structure

In a research project on afforestation for soil and water conservation and its benefit in the Zhonggou watershed, the efficiency of different soil and water conservation methods was studied. The watershed has a sub-humid climate but is regularly subjected to drought. The annual rainfall is 553 mm, of which 66 per cent falls between June and September. The locust tree was selected as the main variety owing to its high adaptability to different environments and resistance to adverse conditions.

The tall vegetation cover in the watershed before the project began was mostly scattered locust trees. Shrubs and grasses hardly grew under the tree canopies. Locust tree leaves are small and easily blown away after falling, leaving little ground cover and resulting in severe soil erosion during heavy rain. The runoff and soil erosion rates for different vegetation structures based on the results of a nine-year monitoring study are shown in Table 7.2.

It can be seen that the runoff reduction for different tree structures is at about the same magnitude, 88–93 per cent when compared to bare land. The results for erosion control however are quite different. On the land with single trees, soil erosion reduces to 1.7 per cent of that on bare land, but compared to the land with sparse woods, it is still more than 15 times as great. This is because, in sparse woods, the light and micro-climate conditions are better than land under single trees, as it allows the growth of shrubs and grasses. The fallen leaves from the shrubs and trees form a complete ground cover. With this kind of multi-layer structure, soil erosion is less than 3 g/m² per year. The erosion reduction amounts to over 99 per cent.

An economical way was developed to form the mixed structure found under a sparse locust tree woodland. Originally, locust trees were planted at a density of 7,000 plants/ha. The survival rate was estimated at 70 per cent, with the number of the surviving trees around 5,000/ha. After five to seven years, when the trees had grown to the size of a rafter,[14] a 50 per cent thinning was undertaken. This provided better conditions for faster growth of the remaining trees. Another purpose was to provide space for the growing of shrubs. The stump of the cut locust was allowed to grow into a shrub through a form of coppicing so that a mixed structure of locust trees and shrubs plus grass resulted. After shrub growth of three to four years, a healthy central branch remained and the other branches were pruned so that the shrub could grow into a mature tree. When the locust trees remaining from the original planting had grown up to a size of a purlin[15] in a 12-year period, a second intermediate cutting was taken and again the stumps of the trees used for growing locust shrubs. Meanwhile, the locust shrubs from the first pruning had grown into trees and after two more years were cut as rafter timber. The trees and shrubs changed roles after each pruning and the mixed structure was maintained. In

Table 7.2 Runoff and soil erosion rates for different vegetation structures

Type of tree structure	Bare land	Land under single tree	Grassland	Land under sparse woods
Annual runoff (litres/m²)	38.9	4.45	2.93	4.68
Annual soil erosion (g/m²)	2,583	45	8	<3

a 12-year period about 3,300 rafters and 1,700 purlins for house construction can be produced on every hectare and at the same time soil erosion can be effectively prevented.

Another way of forming a mixed structure of trees, shrubs and grasses is to plant mixed forests of conifer and broadleaf trees. The locust is planted at an interval of 1 m in rows 2 m apart. Between every two rows of locust trees, a single row of larch trees can be planted. Every year the locust trees are pruned to ensure enough light for growth of the larch trees and the grass. After five years, when the locust trees grow to the size of a rafter, they are coppiced and grown as bushes. This mixture of conifers and deciduous trees provides a more productive form of management.

Role of trees and grass in soil erosion control

Both trees and grasses have a significant role in retaining runoff and thereby reducing soil erosion. In the study in Baozigou watershed from 1988 to 1992, rainfall, runoff, and soil erosion were monitored at different sites. The results showed that over this five-year period the tree-covered land retained and consumed (through evapotranspiration) 75 per cent of rainwater received and reduced soil erosion by over 66 per cent, while land covered by shrubs retained and consumed 76 per cent of rainwater runoff and prevented 47 per cent of soil erosion.

In the Zhongguo watershed, after re-vegetation, the area of forest increased from less than 6 ha (<3 per cent) of the basin to 76 ha (37 per cent of the basin). Soil erosion reduced from 8,835 t/km^2 annually to only 307 t/km^2 (3.5 per cent of the original). Monitoring showed that 92 per cent of soil erosion was from bare slopes with an area of 13 ha, just 6 per cent of the basin. Slope land planted with trees and grass had an area of over 97 ha, 48 per cent of the basin, yet produced only 8 per cent of the total soil erosion.

Bio-engineering measures for soil erosion control

Erosion in gullies and along the steep slopes of valleys (see Figure 7.2) contributes greatly to land degradation in watersheds. Both runoff erosion and gravitational erosion are significant in these locations. The latter type of erosion consists of mass sliding and collapse along the valley/gully banks. A survey in the Zhonggou watershed showed that sliding accounted for 78 per cent of the total gravitational erosion. This can be divided into slipping of surface soil, shallow sliding, and sliding plus collapse. Collapse occurs when a wide vertical crack occurs parallel with the cliff face, causing a large piece of earth to fall.

Gravitational erosion is affected by both external and internal factors. The main external factor is rainfall. Runoff during heavy storms scours the toe of the bank. Rainwater also infiltrates into fissures and cracks in the soil, eventually producing sliding and collapse. Freeze-thaw action due to temperatures fluctuating above and below zero in the winter can also cause collapse and sliding. Internal factors include the shearing strength of the soil, the height and slope of the gully bank, and the elevation of the base level of erosion. The soil shearing strength depends mainly on the density and moisture content of the soil. With higher density and less moisture, the soil has a higher strength. Greater bank height and a lower erosion datum plane aggravate erosion.

In Gansu, measures to control erosion in gullies and on steep valley banks and to reduce sediment discharge include planting trees and grasses, and building check dams and silt retaining dams (creating a flat surface upstream) to retain sediment within the watershed.

Planting trees and grasses on the banks of the gully or valley has been described in previous sections. The following section introduces measures for preventing gravitational erosion and building dams in gullies and valleys.

Measures for preventing gravitational erosion

Measures for gravitational erosion control include:

- Divert runoff away from locations where land slips occur, such as the edge of gullies or the tops of steep slopes. Where slips have occurred planting trees and grass stabilizes the slope.
- Avoid terracing very steep slopes. Because terraced land can retain more rainfall, this results in increased soil moisture storage, weakening soil strength, and increasing the risk of land slippage.
- Plant with care, as while the root systems of trees can help consolidate the soil deep roots can also enlarge fissures and cracks, in loess soils. As a result, water can seep in and widen cracks causing eventual disintegration. In situations where the soil layer is thin and the slope steep, caution should be exercised before planting trees.

The bio-engineering measures in gully and valley bottoms

For wide valley bottoms, contour strips have been built and poplar, willow, and locust trees planted. The tree belt prevents scouring of the valley floor. In narrow deep gullies, to prevent erosion of the gully floor and to reduce sediment discharge and control flooding, dams have been built. Three kinds of dams are commonly constructed in Gansu watershed management schemes.

Check dams. The check dam is designed to retain sediment behind it while allowing most of the runoff to pass through. Check dams are generally built in a cascade sequence in gullies with gradients of 5–10 per cent. This arrangement of check dams is shown in Figure 7.5. The level of the dam crest is about the same as the level of the base of the next dam upslope.

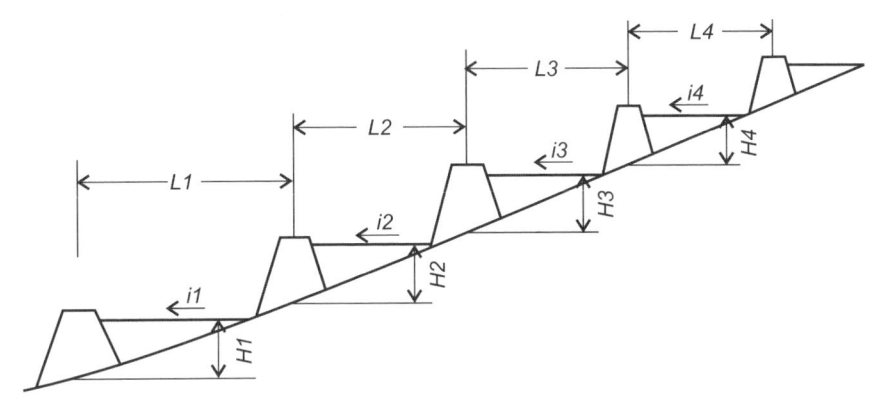

Figure 7.5 Cascade group of check dams

The check dam height ranges between 2 m and 5 m. The dam crest should have the same level as the top of the gully trough or be 0.5–1 m higher. Depending on the availability of building materials, the check dam can be built of willow tree posts, willow branches, earth, or stone. In the loess area of Gansu, check dams built of stone are rare as there is a general lack of rock.

Check dams made of willow are composed of live willow posts, mats made of willow branches and stone. The posts are 1–1.8 m long and 5–8 cm in diameter with a sharpened end. They are inserted into the ground 0.5 m deep in two rows with an interval of 0.5 m between rows and piles. The maximum height for the willow dam is 3 m but they are usually less than 2 m high.

The branches are woven into mats and attached to the posts and gravel is filled in between the rows. Runoff water can pass through these porous dams but most of the silt is retained. Branches can grow up to 1 m long from the live willow pile in a year. Photo 7.3 shows a willow check dam.

Check dams made of earth have a height of 2–5 m and crest width of about 1.5–2 m. The gradient of the dam slope ranges between 1:1 and 1:1.8 (vertical to horizontal). The sizes of earth check dams with different heights are listed in Table 7.3.

A spillway should be built on bedrock or firm ground by the side of the gully bank to release floodwaters. A typical flood with 3–6 hr storm rainfall of 10–20 year recurrence interval is analysed. The shape of the spillway can be trapezoidal or rectangular.

Since the design of spillways is a critical feature of any dam project, care should be taken over the structural design, which ideally should be based on sound hydrological, geological, and structural analysis.

Silt-retaining dams. The silt-retaining dam is designed to retain sediment and at the same time to form flat land upstream, which will eventually be cultivated. Photos 7.4 and 7.5 show newly built silt-retaining dams and the reclaimed land behind an old series of dams.

116 EVERY LAST DROP

Photo 7.3 Willow check dam

Table 7.3 Size of earth check dams

Dam height (m)	Top width (m)	Bottom width (m)	Upstream slope gradient	Downstream slope gradient
2	1.5	5.9	1:1.2	1:1.0
3	1.5	9.0	1:1.3	1:1.2
4	2.0	13.2	1:1.5	1:1.3
5	2.0	18.5	1:1.8	1:1.5

In China, silt-retaining dams are classified into large, medium, and small sizes according to the dam height, storage capacity, and area of the reclaimed land behind the dam (see Table 7.4).

Silt-retaining dams are built along gullies with floor slopes of less than 3 per cent. For small gullies with catchment areas less than 1 km^2, the first silt-retaining dam should be built at the downstream end. When this has silted up and the reclaimed land planted with crops, then a second one is built upstream to protect the land from flooding. Then a third one can be built and so on.

For catchment areas of 3–10 km^2, construction of dams should start upstream and proceed downstream, with the height and storage capacity

Photo 7.4 Newly built silt-retaining dam

Photo 7.5 Reclaimed land behind a dam

increasing successively downstream. Another procedure is to build dams at the lower end and middle of the stream simultaneously. After they are silted up, the dam heights will be increased and at the same time a large or medium-size reservoir should be built upstream for flood control to protect the reclaimed land. For catchment areas of 10 km² or more, dams should be built first on the tributaries upstream of the main channel and then the dams in the lower catchment, with appropriately larger capacities.

It is important to take measures to protect reclaimed land formed behind the dam. After the dam has silted up and the land reclaimed, flooding needs to be controlled by building dams upstream. If the reclaimed land is located in the main gully then dams for flood control should be built on the tributaries upstream. To protect the land from flooding, a protection dike needs to be built so floodwater is drained along a ditch to the spillway of the next downstream dam.

The sediment deposited behind the dam is nutrient-rich, fertile, and contains high soil moisture. A good yield can therefore be expected. In the Nanxiaohegou watershed in Xifeng Municipality of Gansu, 11 silt-retaining dams were built, reclaiming 6.7 ha of eroded land. Monitoring in 1978 revealed the 11 dams had retained 36 t of total phosphorus, 150 t of total nitrogen, and 618 t of organic matter.

The silt-retaining dam is composed of the embankment, spillway, and/or flood discharge culvert. In the *China Regulation of Techniques for Comprehensive Control of Soil Erosion* (China Bureau for Technical Supervision, 1996), the design standards for flood control are stipulated, as shown in Table 7.5.

Table 7.4 Classification of silt-retaining dams in China

Dam classification	Dam height (m)	Storage capacity (10^3 m³)	Area of land behind dam (ha)
Large	>25	500–1,000	>7
Medium	15–25	100–500	2–7
Small	5–15	10–100	0.2–2

Table 7.5 Flood control standard for silt-retaining dams

Item		Flood control standard for dams with volume (10 m³)			
		<100	100–500	500–1,000	1,000–5,000
Flood recurrence	Design flood	10–20	20–30	30–50	30–50
interval (year)	Check flood	30	50	50–100	100–300
Period of silt-up (years)		5	5–10	10–20	20–30

Reservoirs. In locations where there are springs or other water sources, small reservoirs can be built for either irrigation or aquaculture. These can also act as buffers to reduce the peak flood flows to protect downstream dams. Silt-retaining dams should be constructed upstream of a reservoir to prevent loss of storage from siltation.

Dams play an important role in decreasing soil erosion and reducing the sediment discharge from watersheds. With deposition behind check dams and silt-retaining dams, the datum plane can be raised so the length of the slopes of both banks, as well as the longitudinal slope of the gully, are reduced. Since most of the silt from the whole watershed is retained behind the dams, sedimentation downstream is mitigated.

Impacts of small watershed management in Gansu

Practical experience over the past 30-plus years has shown that the effects of small watershed management on rural development and environmental conservation have been significant. There have also been downstream benefits for the whole river basin from reduced silt discharge. However, measures taken through watershed management also reduce outflows, which bring about a reduction of the runoff in the downstream river basin. This has resulted in the reduction of flow in the Yellow River. The following sections provide an assessment of the impact of watershed management in Gansu.

Impacts of watershed management on ecosystems and micro-climate

One of the main measures taken in watershed management in Gansu is afforestation and grass planting. Before intervention, vegetation coverage was very low, usually lower than 10 per cent. Furthermore, the farmers dug up the residual straw and grass for use as fuel, thus removing all vegetation. During watershed management, the area of tree and grass coverage is greatly increased. For example, in the Zhonggou watershed, within five years, there were 71 ha of trees and 16 ha of grasses planted. The newly vegetated area accounted for 40 per cent of the total watershed. In Qianjiagou watershed in 1983–7, new forest and grassland increased by 6 km² to cover 11 per cent of the watershed. In Jinchuan County, there were 428 km² of forest and grass planted in 1991–5. The new vegetation area amounted to over 30 per cent of the total territory. Increase of greenery not only benefits control of soil erosion but also improves the local climate. Observations in Zhonggou watershed revealed that after shelter trees were planted on terraced land, the humidity increased by 4 per cent, average wind speed decreased by 48 per cent, and the soil moisture increased by 10 per cent. Furthermore, the period of exposure to high temperatures and low humidity caused by dry hot winds in summer was shortened by 3.3 and 4.6 hours, respectively.

Downstream impacts of watershed management

Reduction of silt discharge. The most obvious effect of the watershed management is soil erosion control. Before watershed management interventions, annual soil erosion in the loess areas in Gansu ranged from 5,000 to 10,000 t/km^2. Since the watershed management interventions described above were fully established, soil erosion on most of the land is now under control. As a result, silt discharge from the watershed has greatly reduced. Table 7.6 shows a reduction of the soil erosion and silt discharge versus the coverage of erosion control measures (the ratio between the area where the soil erosion has been controlled and the total land).

Although the area of each watershed is small compared to the whole Yellow River Basin, the number of watersheds is large. In Gansu, more than 600 small watersheds with a total area of 12,000 km^2 (more than 10 per cent of the loess area) now have soil conservation measures in place. Watershed management has had a major impact on the hydrology of the whole river basin.

A study was made in 1996 to assess the impacts of soil and water conservation on the runoff and silt discharge of Zuli River, one of the large tributaries of the Yellow River in the loess area. The Zuli River Basin has an area of 10,653 km^2 and annual precipitation of 376 mm. In previous decades, soil and water conservation measures were introduced on 2,403 km^2 (23 per cent of the river basin). Using 1955–69 as the baseline, the silt discharge in 1970–9 reduced by 25 per cent, and in 1980–9 by 37 per cent. After an analysis using hydrological models, reasons for the reduction were divided into those resulting

Table 7.6 Reduction of silt discharge versus coverage of control

Name of watershed	Area (km^2)	Period before control	Period after control	Coverage of control (%) Before	After	Runoff (m^3/km^2) Before	After	Silt discharge (t/km^2) Before	After
Zhonggou	2	1979	1992	4.9*	47.3*	65,100	11,627	8,885	60
Qianjiagou	54	1982	1987	48.9	81.9	45,000	11,700	8,000	1,820
Baozigou	18	1987	1992	29.7	86.6	84,916	34,000	10,279	2,206
Jinchuan county	1,409	1972	1993–5	16	78	83,135	43,005	7,974	2,691

*Note:** Denotes the coverage of only forest and grassland (terracing not included) in the total area.

from natural causes and those due to human activity. The main natural cause was due to changes in the amount and pattern of precipitation. The impacts of human activities included soil and water conservation measures and silt deposition behind dams. The reduction of silt discharge in the 1970s and 1980s due to natural causes was 2 per cent and 12 per cent, while that due to human activity was 23 per cent and 25 per cent, respectively. Around 1.6 billion tonnes of silt enter the Yellow River each year, mostly from the loess areas. Reduction of silt discharge from watersheds into the downstream river basin is very important to mitigate the sedimentation problem of the Yellow River.

Mitigation of flooding problems downstream. Since the general adoption of watershed management measures, most of the rainfall is now retained in the soil. Part of it becomes soil moisture and the rest becomes groundwater flow, which eventually returns to the river system. As a result, the peak flow in the stream is sharply reduced and most of the yearly runoff is in the form of base flow. Therefore flooding problems downstream have been greatly reduced. Table 7.7 gives a comparison of peak flood flow during a heavy storm between Qianjiagou and the neighbouring Tianjiagou watershed, the areas of these two watersheds being very close to each other. Both watersheds encountered a similar storm in 1987 but the result was distinctly different in each. In Qianjiagou watershed, 81 per cent of the area was subject to erosion control. In the storm, the peak discharge was only 14 per cent of that occurring in Tianjaigou watershed, where the soil erosion control measures had only been partially introduced.

The composition of runoff in the watershed has also changed dramatically since soil conservation measures were introduced. In Baozigou watershed, before erosion control in 1987, the yearly rainfall and runoff was 544 mm, with almost all the runoff in the form of floodwater. After watershed management took place in 1988–92, the mean yearly rainfall and runoff was slightly less at 540 mm, but floodwaters accounted for only a third of the yearly runoff, with the remainder in the form of base flow.

Reduction of runoff downstream. The effect of terracing, land levelling, and afforestation has been a significant increase in the amount of rainwater retention by the soil, thus greatly reducing the outflow from watersheds. Table 7.8

Table 7.7 Comparison of flood discharge between Qiangjiagou and Tianjiagou watersheds

	Area (km^2)	Control (%)	Rain (mm)	Intensity (mm/min)	Peak flow (m^3/s)	Flood amount (10^{3}m^3)	Sediment discharge (t)	Sediment module (t/km^2)
Tianjiagou	58	30	53.2	0.043	22.5	251	108,000	1,860
Qianjiagou	60.58	81.9	52.1	0.95	3.04	3.08	4,930	81.4
Reduction					-86.5%	-88.2%	-95.6%	

Table 7.8 Runoff reduction versus percentage of erosion control and silt reduction

Name of watershed	Period of harnessing	Percentage of control Start of period	Percentage of control End of period	Runoff reduction (%)	Silt discharge reduction (%)
Zhonggou	1979–92	5*	47*	82	99
Qianjiagou	1982–7	49	82	74	77
Baozigou	1987–92	30	87	60	79
Jinchuan County	1972 to 1993–5	16	78	48	66
Zuli River Basin	1955–89	0.5	22	11**	19**

Note: * Denotes the coverage of forest and grassland over the total area. ** Reduction percentage comparing the average value of 1980–9 with that of 1955–69.

shows the runoff reduction related to the percentage of erosion control and the reduction of silt discharge in five watersheds.

From Table 7.8, it can be seen that the outflow of watersheds was reduced after the introduction of soil control measures in small watersheds by 60–82 per cent. In the case of the Zuli River Basin, a runoff reduction of 11 per cent resulted after 20 years of soil and water conservation projects. This clearly shows the impact of watershed management on runoff reduction downstream, impacting the whole Yellow River Basin.

Some experts view runoff reduction downstream as a negative impact of watershed management in the loess areas of the Yellow River Basin and one of the causes of the reduced downstream flow of the Yellow River. However, when the impacts are considered in an integrated way, another perspective emerges. First, the small watersheds in the loess-covered region are among China's most water-scarce areas. There is little or no surface water or groundwater available. Rainwater is the only available source and for any development to take place its efficient use is essential. Second, the runoff coefficient before watershed management schemes were introduced in Gansu's loess area was only a round 0.1 and it was estimated that each person shared only around 300 m^3 per year, much lower than the mean water resource availability in China as a whole. Third, although the runoff from the watershed reduced, sediment discharge also decreased, even more significantly. Data in Table 7.6 show that every cubic metre of runoff reduction is associated with 0.13–0.19 t of silt reduction, meaning each tonne of silt reduction equates to 5.4–7.6 m^3 of water retained. This number indicates the high efficiency of sediment management methods for the lower catchment. For example, at the huge Xiaolangdi Reservoir located downstream on the Yellow River, 10 experiments on the water and silt regulation were conducted from 2002 onwards. Between 2002 and 2009, a total of 33.1 billion m^3 of water had to be drained to flush out 575 million tonnes of silt from the reservoir and the riverbed into the sea. The drainage of each tonne of silt thus required almost 58 m^3 of water. Overall the comprehensive impacts of watershed management on the downstream river basin appear to be positive.

Enhanced agricultural production and improved living standards

Before the introduction of watershed management, 70 per cent of cultivated land was sloping. The depletion of top soil and the resulting nutrient loss every year kept productivity and crop yields at a low level. Terracing and land levelling during watershed management reduced soil erosion and nutrient loss. This allowed farmers to start to change their cultivation practices from an extensive pattern to an intensive one, resulting in a significant increase in agricultural production. For instance, in Qianjiagou watershed, crop yield on levelled and terraced land has increased by 890 kg/ha and 1,245 kg/ha, respectively. In Baozigou watershed, yield averaged 2,850 kg/ha in the period of 1988–92 after soil and water conservation measures were introduced, compared to only 1,350 kg/ha in 1987. The production of the cropping, husbandry, and forestry sectors increased around three-fold. Incomes increased by more than 2.5 times and annual grain production increased from 267 kg to 545 kg per capita.

The impact of afforestation was also significant. Villagers collected 2,700 tonnes of wood and straw and effectively met their demand for cooking fuel. In Jiuhuagou watershed, when the terracing and afforestation was complete, there was a major diversification of agricultural production. In a four-year period, 67 ha of fruit trees were planted, including Chinese prickly ash, pear, and apricot, and 1,500 rainwater tanks were built for both domestic water supply and for the irrigation of 207 ha of land for cash crops. Furthermore 96 households began pig breeding and an area of 133 ha was set up for producing seed and commercial potato crops. Per capita incomes increased by 97 per cent, and grain production per capita increased from 410 kg to 654 kg.

It is interesting to consider the economics of watershed management. What would be the optimum division between terracing and afforestation in watershed management of the Gansu loess area? In the early 1990s, an economic evaluation of watershed management in the province concluded that terracing is economically feasible. At that time, terracing in the Gansu loess area was only done manually as labour had a very low value, especially in areas where the market economy was not yet well established. Terracing could pay for the large amount of low-cost labour used due to the considerable benefit of increased agricultural productivity. This, however, applied only where the upper limit of the slope was 25°. Land steeper than this was better shifted from cultivation to the planting of trees and grass. Now that the market economy has developed in China, even in the loess areas of Gansu, labour is far more costly than in the past. Investigations have shown that both afforestation and grass planting can have a similar impact to terracing in terms of erosion control. For example, in Zhonggou watershed, new forest and grassland reduced soil erosion by as much 99 per cent when compared to bare sloping land. In Qianjiagou and Baozigou watersheds, forest and grassland similarly reduced soil erosion by 80 per cent and 75 per cent respectively.

Box 7.3 Case study: The life-changing impact of watershed management

Ms Zhang Yun married Mr Wang Tingsheng, a farmer in Jiuhuagou village, in 1970. When she first came to the village she found life was so hard that she thought of leaving her husband and returning to her native village. She had to spend more than three hours every day fetching water from the bottom of a gully. Sometimes she went to a neighbouring village to 'steal' spring water but often was stopped by the local inhabitants. Her fields yielded so little that her family of four usually had only 300 kg of grain per year, which was far from enough.

Now her family has four water cellars with volumes of 40 m^3 each. The water is more than enough for her family. She has participated in the Land Conversion Programme by shifting 0.7 ha of land to planting alfalfa and cypress trees. For this she gets $300 of compensation each year from the government. Her original 1.1 ha of sloping land was terraced between the 1970s and 1990s. It was very hard to make the changes but now she enjoys the results as yield has increased by more than 30 per cent.

She harvests more than 4,000 kg of wheat, corn, and oil seeds. She has 0.27 ha of potato and a small orchard producing pears. She also raises pigs. Her husband is a carpenter and finds temporary work in the township. Her life has been dramatically changed, mainly she thinks from the benefits of terracing and RWH.

Photo 7.6 Zhang Yun's home

PART IV
Challenges, future prospects, and conclusions

CHAPTER 8
Challenges and prospects for rainwater harvesting in Gansu Province

Although much progress has been made in RWH developments in Gansu over the past 20 years, many challenges remain. An evaluation conducted in 2005, while confirming the significant achievements of the programme, also revealed some major challenges.

Challenges of rainwater harvesting for domestic water supply

The Chinese Ministry of Water Resources criteria for safe rural water supply specify a daily quota of 20–40 litres per person in semi-arid areas, with a supply reliability of 90–95 per cent. The water supplied also needs to meet the national Standards for Drinking Water Quality (GB 5749—2006). Meeting these criteria presented several challenges for the RWH programme for domestic water supply.

Water demand

In the Loess Plateau region of Gansu, the mean annual rainfall ranges from 250 mm in the north to 550 mm in the east, with most areas receiving around 400 mm. To ensure requirements for water demand and reliability were met, catchment areas (including roofs and concrete-lined courtyards) needed to range from 230 to 440 m^2. In most villages, the area of roofs plus courtyards was less than this, resulting in the daily water supply quota falling short of the stipulated amount. Through the New Village Construction Programme, since the early 2000s houses have been upgraded and living conditions improved through several pilot projects, for example in Daping village in Qingran township. In these cases, households have generally been reaching the required standards regarding the quantity and reliability of rainwater supplies. To meet the demand for the whole region, however, significant additional government inputs along with farmer contributions (in the form of labour and cash) will be necessary.

Water quality

Owing to the dry climate in Gansu, roof and ground catchment surfaces are rarely washed clean by heavy rain. As a result, the quality of the stored rainwater generally cannot meet the national standards in terms of turbidity and biological indices, the upper limits of which are often greatly exceeded. Reasons for poor water quality in water cellars include:

- Due to the absence of gutters and downpipes, the runoff from roof and concrete surface catchments is contaminated during contact with the ground.
- First-flush systems for enhancing water quality are not generally accepted by local farmers.
- Key design features to protect water quality are absent, such as screens over inlets and overflows etc.
- Regular water quality monitoring is not being undertaken.

Despite the poor quality of the water in the cellars, some farmers consume this directly without boiling or treating it in any way, although in most cases the water is boiled using the solar cookers described in Chapter 3. It is also interesting to note that while bacteriological water quality may not generally meet the required standards, there have not been any significant cases of disease outbreaks or health problems directly attributed to the direct consumption of untreated rainwater in Gansu Province. Nevertheless, a number of measures are recommended that would greatly assist in improving the quality of rainwater stored in the cellars and mitigate any potential health problems. These include:

- Develop a standard design for rural RWH systems for domestic use based on input from relevant government agencies together with community inputs.
- Encourage the use of a dedicated water tank or cellar connected to a roof catchment with properly designed guttering and downpipes for storing rainwater for drinking, cooking, and washing.
- Explore the dissemination of effective and affordable filtration equipment to further improve the quality of rainwater for potable supplies.
- Promote health and hygiene education, particularly relating to the safe collection, storage, and if necessary treatment of rainwater supplies for domestic use.

Challenges of supplementary irrigation with rainwater harvesting

The challenges of providing additional irrigation using RWH systems include:

- Analysis of data collected from 1995 to 2004 on mean catchment area and runoff coefficients revealed that the rainwater harvested can only fill about 39 per cent of the available storage capacity of the water cellars. Looked at another way, this suggests almost two-thirds of the available rainwater tank volume is effectively not being used.

    Since most of the RWH projects were constructed by the farmers themselves it is clear that while they have paid a lot of attention to

building the storage tanks, less consideration and planning have been given to the catchment areas. It seems that the farmers have been less aware of the importance of the adequacy of catchment areas as an essential component of the RWH system. This shortcoming reveals a weakness in the level of technical advice and support given by RWH projects and agricultural extension services.
- Another underutilized aspect of the RWH programme in Gansu is irrigation potential. Investigations have shown that of the 80,550 ha of land that is capable of being irrigated from rainwater cellars, the total area (including greenhouses) that is equipped with drip, sprinkler, or micro-spray irrigation systems only accounts for 38,620 ha, 48 per cent of the total. Most of the remaining 52 per cent of land is partially irrigated but only uses simple manual methods that are labour intensive and inconvenient. The main reason for this is lack of funding. Most of the farmers in the area are still poor and cannot afford to buy modern irrigation equipment or access loans for such purchases.
- The main obstacle preventing more widespread adoption of RWH-based irrigation in Gansu is a lack of awareness among farmers. Since the farmers in this region have been used to cultivating rain-fed crops, they would normally only consider using supplementary irrigation if they were subjected to a serious drought. One of the authors has visited RWH irrigation projects several times and found water cellars full of rainwater, while the surrounding cultivated land was not being irrigated. Among the reasons given for not using the stored rainwater were: (1) lack of necessary equipment such as pumps; (2) the view that irrigation is only necessary to ensure crop survival but not as a means to raise crop yield and/or value; (3) lack of knowledge about how to irrigate; (4) lack of motivation, especially where there is no market for cash crops in remote locations, often due to poor transport infrastructure.
- Although the government has paid attention to the construction of the RWH systems for both domestic purposes and irrigation, less attention has been given to the capacity and awareness-building of the farmers. While funding is often available for building the RWH irrigation systems, it is harder to access support to train farmers to fully understand how to maximize the use of these systems and utilize valuable rainwater as efficiently as possible. There are many examples of successful experiences of converting stored rainwater into food, cash crops, and money, yet even in Gansu, most of the farmers are still not fully aware of either the potential or opportunities arising from the optimal use of RWH technologies.

Prospects for the future of rainwater harvesting

In 2005, to ensure the quality of rural water supplies, the Ministry of Water Resources decided to adopt a policy of providing centralized water supply schemes wherever possible. These water supply systems are to be planned and built for both urban and rural areas in an integrated way. According to the Criteria for Safe Rural Water Supply (China Ministry of Water Resources & China Ministry of Health 2004), the water supply of 160 million rural people, or about 20 per cent of the total, did not meet all the required water quality standards. In Gansu, water supplies of almost 13 million rural people (62 per cent of the rural total) did not meet all the criteria. According to official data, at the end of 2008, over 9 million rural dwellers in Gansu still had substandard water supplies. It is planned to update the water supply for Gansu's rural population to a safe level by the end of 2013 through building a number of centralized water supply projects. Among the remaining rural population, new RWH projects are planned for about 162,713 people living the most remote areas, or only about 1.7 per cent of the total. Despite its unprecedented success in improving domestic water supply and boosting productivity through supplementary irrigation, officials seem to have less confidence in decentralized solutions, including RWH projects, for ensuring a safe domestic water supply.

Some experts, however, still believe the role of RWH in rural domestic water supply should not be underestimated. First, the water quality of the RWH projects can be greatly improved if systems are designed and constructed to a high standard and regular monitoring is undertaken. Several household-based rural water treatment technologies, such as water filters, that can ensure that the quality of rainwater after treatment will fully meet the national standards for drinking water, have recently been developed (see example in Appendix 1). With the ongoing improvement in the rural economy, these filters are becoming increasingly affordable for most households.

Second, owing to the current low price of water, many water supply plants cannot cover even their operation and maintenance costs, let alone depreciation costs. However, raising the price of water will be deeply unpopular among the rural population, especially in the less-developed areas. It therefore seems likely that even if households get access to piped water supplies, they will still use or partially use rainwater since this source is free of charge.

Regarding the utilization of rainwater for irrigation, although at present only a small percentage of land is subject to supplementary irrigation using RWH systems, there have been many successful examples of applying this technology. These have all demonstrated that irrigation using RWH systems can enhance crop productivity significantly, even when using relatively small quantities of water.

On 31 December 2010, the Central Committee of China Communist Party and the State Council made a united resolution on 'Accelerating the Reform and Development of Water Resources'. With regard to addressing the water shortage problem, the resolution seeks 'to support farmers to build the small and mini-size water resources project, to significantly enhance utilization of the rainwater and storm water'. In another clause dealing with low water availability it seeks to 'pay high attention to utilization of rainwater'. The resolution expresses the serious concern of the highest authorities in China regarding the need for rainwater utilization. We can conclude that the development of RWH in China especially in the semi-arid areas such as Gansu should be further strengthened for both the domestic water supply and agriculture production.

In China, rain-fed agriculture covers about 51 per cent of all cultivated land, but in Gansu Province more than 70 per cent is rain-fed. The potential for further expansion of RWH-based irrigation systems in the province is significant. Based on the experiences of the last 20 years, irrigation using RWH-based systems has proved to be a highly effective and innovative way to manage scarce water resources, boost food production, and drought-proof rural communities. It combines conventional dry farming methods with low-rate targeted irrigation for crops at critical periods in the growing cycle. The formula for enhancing dryland farming that has been adopted by authorities and technical people alike is simply:

upgraded dryland farming = RWH + conventional farming practices

All the current indicators suggest that RWH-based irrigation systems provide the key to leading the rural population of the arid Loess Plateau regions of both Gansu and surrounding provinces out of poverty and to becoming a prosperous rural society.

Conclusions

Technology has the potential to greatly improve humanity's quality of life by providing affordable and convenient access to all the basic necessities of life: food, water, energy, and shelter. It is, however, a double-edged sword that can also deliver great harm through facilitating over-consumption and over-exploitation of resources, leading to land degradation, pollution, and climate change.

The inspirational programme outlined in this book has shown how the application of RWH and other simple but appropriate and low-impact technologies, such as solar cookers and low-cost greenhouses, have transformed millions of lives across a large region of rural China. The approaches outlined follow the 'Small is Beautiful' philosophy of E.F. Schumacher (1973) and provide indisputable evidence of the effectiveness of this approach to promoting both human development and environmental sustainability.

Over the past two centuries, humanity has sought solutions to feeding the planet's rapidly growing population and meeting its ever-increasing demand for consumer comforts, generally through applying modern technology at ever-larger scales. In the short term, the benefits of many of the resulting mega-projects, whether they be giant dams taming mighty rivers or nuclear power stations to fuel ever-growing industrial output and consumption, may seem obvious. Nevertheless, their costs in economic, social, and environmental terms have been substantial, their long-term impacts uncertain and their sustainability problematic.

Finally, we currently live at a time when the world is globalizing and everyone is starting to become increasingly aware of the greatest challenge ever to face human civilization, namely that of runaway global warming and all that this implies in terms of greater frequency of floods, droughts, fires, storms, and sea-level rise. It is also becoming increasingly clear that the large-scale energy and resource-intensive technologies and practices of the past are not going to help us to address the challenge, as it is these approaches that have created the problem. Instead what are needed are smart and appropriate technologies that use energy and resources sparingly and operate at a human scale. These not only have the potential to address our problems globally but also offer the possibility of bringing back a sense of reality and control over technology to the individual user.

Clearly high technology, such as the prudent application of computers, and satellite monitoring, will play an important role in the future. Nevertheless, humanity should also re-examine some of the approaches and ancient wisdom from the past and consider their relevance to the issues facing the world today. To face and overcome the challenges of the future, humanity will be best served by combining the most appropriate technologies and resource management strategies of the present with the accumulated experience and wisdom from the past.

It is still not too late to meet the great challenges of our time, 'to make poverty history', and to establish a harmonious coexistence between humanity and the planet, but time is not on our side and the time to act is now.

APPENDIX 1
Water purification system for harvested rainwater in Gansu

The quality of rural water supply has been given increasing attention recently because of the goal of central government to modernize and establish a prosperous society in the whole of China. In 2005, the Ministry of Water Resources issued new higher water quality criteria to ensure the safety of domestic supplies in rural areas. Resulting water quality testing revealed that while most chemical and toxicology results for rainwater samples from water cellars were within national standards, results for biological indices and turbidity to test for possible bacteriological contamination significantly exceeded the standards.

While householders have adopted measures to improve the quality of cellar water, such as placing water cellars inside courtyards, avoiding contact with animals, and cleaning the catchment area before rainfall events, these have been insufficient to ensure good-quality water.

Surveys also found that although solar cookers were present in more than half of the households, and these could produce hygienic water for cooking and tea-making, farmers do not always drink tea, but often drink the water untreated from storage jars kept in the kitchen, where the rainwater from the cellar is stored to allow larger suspended particles to settle. Despite an absence of reports of any waterborne disease outbreaks linked to drinking this water, there is some concern that the water could in the long run be harmful to the people's health. It was thus decided to develop a system using effective and affordable equipment for treating drinking water collected from RWH systems in rural areas.

The 'Research and Demonstration Project on the Appropriate Technique for Securing the Drinking Quality of the Harvested Rainwater in the Northwest Villages' was therefore sponsored by the China Ministry of Science and Technology and undertaken by GRIWAC between 2008 and 2010. The project site was located at Caijiamen and Zhangjiabao villages in Huining County, where the mean annual precipitation is around 393 mm.

Water was tested from the water cellars that were using three types of catchment: roof plus concrete-lined courtyard, roof and natural earthen slope, and sealed road. The results showed that the turbidity was around three times the permitted limit and total coliforms exceeded the limited by almost 70 times. E. coli (*Escherichia coli*) was 270–302 MPN/100 ml, while the maximum permissible is zero (MPN refers to most probable number and is an estimation of the bacteria concentration based on the probability theory. It is used in measuring the E.coli in the Chinese standard for water quality).

134 EVERY LAST DROP

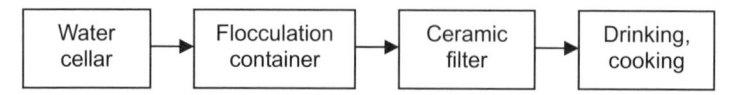

Figure A1.1 Flow chart of the purification system

Since bacteria often attach themselves to suspended particles, reducing turbidity results in a reduction of the presence of bacteria. The effective operation of filtration equipment also requires the removal of suspended particles. The purification system uses a strategy of first reducing the turbidity and then screening out the bacteria. The operation involves the use of a flocculation and sedimentation chamber and filtration system, as shown in Figure A1.1.

For the flocculation process, polyaluminium ferric chloride (PAFC), a macromolecular inorganic coagulant composed of aluminate and ferrite chloride is used. Tests were carried out to compare the sedimentation process of the cellar water with and without adding PAFC and also to find the best concentration of PAFC. For the water with no PAFC added, it took 120 hours for the turbidity to drop from the original value of 18 to 3.6 nephelometric turbidity units (NTUs), which was still higher than the permitted limit of 3 NTU. When PAFC was added, the turbidity dropped to 0.3–1.3 NTU after just 60 hours. A concentration of 40 mg/l of PAFC was found to be the optimum.

Only the water required for drinking should be extracted from the cellar for the process of flocculation. A special cylindrical container has been developed for the flocculation process. It is installed with inflow, outflow, and drainage valves, as well as an additive inlet. Water is pumped from the cellar into the container and a specific amount of PAFC coagulant added according to the volume of water needed. After the settling process is finished, the clean water can flow to the filtration equipment (see Figure A.1.2) by gravity. Sediment at the bottom of the container is drained out periodically. The volume of the

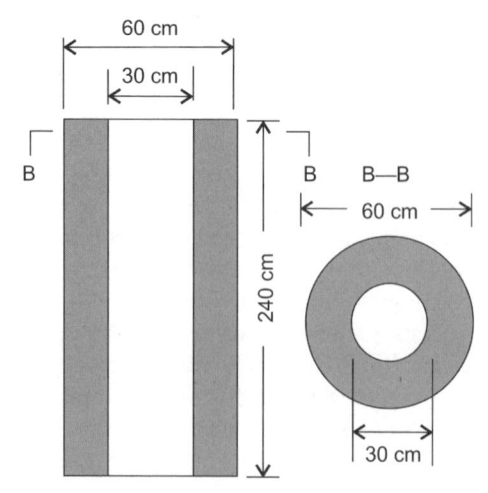

Figure A1.2 Section of the filter

flocculation container depends on the number of family members and daily water demand. Containers with volumes of 200, 250, 300, 400, 500, 800, and 1,000 litres have so far been designed.

In the past, filters could be of polypropylene fabric (1 mm), ceramic (0.1–0.01 mm), and a cation exchange resin. The polypropylene fabric was easily blocked by particles in turbid water and the resin was costly. To overcome this, a purely ceramic filter has been adopted. The ceramic argil is composed of kaolin, diatomite, montmorillonite, palygorskite, and alumina. Experiments were carried out to find the suitable proportion of the components. The sintering temperature is up to 1100°C. The ceramic filter is cylindrical, with an outer diameter of 60 cm and a 15 cm thick wall as shown in Figure A1.2. Water flows from the flocculation container to the ceramic filter by gravity and seeps through the wall into the inner chamber, where an outflow pipe is connected. The wall has openings of 0.1 mm. It can thus screen off very fine suspended particles as well as bacteria. The filter can be used repeatedly. If the surface is blocked by the sediment it can be easily cleaned by rinsing with water or wiping with a clean cloth.

Photo A1.1 The purification system

Photo A1.1 shows the installation of the purification system. It also shows the active carbon filter installed by the side of the ceramic filter. While not essential, this can further purify the water and improve the smell.

The cost for the whole system is about 1,500 CNY ($240). The running cost to purify 1 m^3 of water is only 0.8 CNY($0.13). The service life for the container and the ceramic filter is expected to be 30 years. The active carbon filter needs to be replaced every six months to one year.

The purification system has been installed in 50 households for demonstration purposes in Caijiamen and Zhangchengbao villages of Huining County, paid for by the project. The system is well accepted by the farmers, who appreciate the treated water very much.

There are, however, challenges to wider replication. First, the price of the purification system remains too high for most farmers to buy at present. One solution would be to convince the government to subsidize part of the cost. This might be appealing as the government will have to spend much more if it decides to fund centralized rural water supply projects with expensive piped distribution systems and high running costs. Another solution might be that the price of the systems could be brought down significantly if they were mass produced in large numbers. Second, to date, the proposed widespread implementation of this project has not been agreed to by the Ministry of Science and Technology. The long-term performance and reliability of the systems have yet to be proven. Whether the claimed 30-year life expectancy is realistic without the need to periodically change the ceramic filter is also unknown.

Photo A1.2 An 80-year-old man enjoys the treated water

APPENDIX 2
Solar cooker design details: The parabolic curve and how to draw it

Equation A2.1 is used to derive the parabolic curve used for the cooker:

$$4py = x^2 \qquad (A2.1)$$

In this equation, x and y represent the horizontal and vertical coordinate and p is the focal distance.

To draw a parabolic curve, first we should determine the focal distance. In China the recommended standard focal distances for solar cookers of different sizes are shown in Table A2.1. In the table, the interception area represents the projection area of the cooker perpendicular to the sun's rays when the axis is parallel to the sun's rays. A cooker for a family with four to six members requires an interception area in the range of 1.5–2.0 m².

When the focal distance is determined, the parabolic curve can be drawn as follows (the procedure is also shown in Figure A2.1). Here we take the focal distance p as 70 cm as an example.

Step 1: when p = 70, the interception area may take 2 m² and an estimation of the maximum horizontal coordinate B = (2÷3.14)$^{0.5}$ = 0.8 m. The maximum vertical coordinate h can be calculated by the following equation:

$$h = B \times B \div 4 \div p = 0.8 \times 0.8 \div 4 \div 0.7 = 0.229 \text{m} \qquad (A2.2)$$

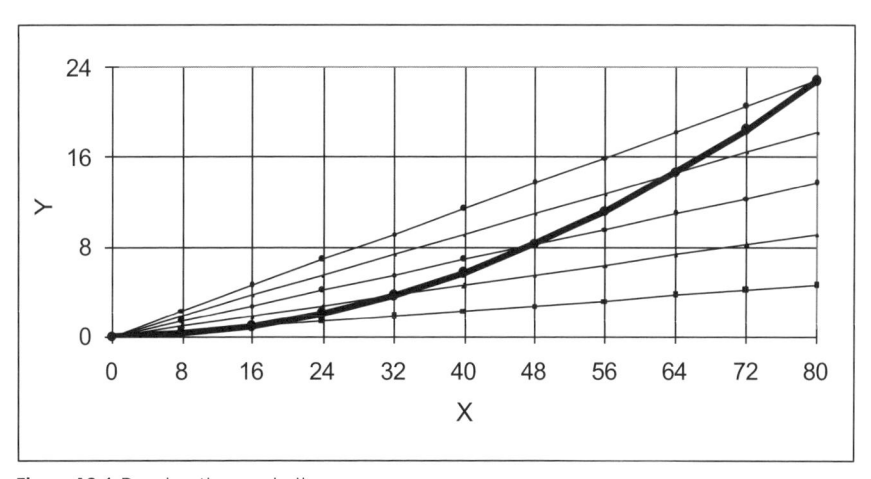

Figure A2.1 Drawing the parabolic curve

Table A2.1 Recommended focal point distances for solar cookers

Interception area (m^2)	1.0		1.2		1.6		2.0		2.5		3.2		4.0	
Focal point distance (cm)	50	55	55	60	60	65	70	75	75	80	85	90	95	100

Source: China Standard for Focusing Solar Cooker (NY/T 219—2003)

Step 2: on a piece of paper with width of 100 cm, draw the vertical and horizontal coordinates with the lengths of 22.9 cm and 80 cm, respectively. Step 3: divide the two coordinates into 10 intervals. Each interval on the vertical and horizontal coordinate is 2.29 cm and 8.0 cm, respectively. Step 4: link inclined lines between the origin and the divided points on the vertical coordinate. Draw vertical lines from each of the divided points on the horizontal coordinate. Step 5: we get 10 intersections between the inclined lines from lower to top in turn and the vertical lines from left to right in turn. Step 6: link all the 10 intersections to form the parabolic curve.

Notes

1. The loess soil, like all other kinds of soil, is composed of clay, silt, and sand, with particle diameters ranging from 0.001 mm to >0.05 mm. Silt (0.005 to 0.05 mm) makes up the main component of loess soil.
2. The 1-2-1 Project was launched in Gansu Province following the severe drought of 1995; it helped provide 1 million people with one catchment composed of a tiled roof and concrete lined courtyard, two rainwater tanks, and one piece of irrigated farmland.
3. The 1-1-2 Project is a RWH-based irrigation project in Inner Mongolia Autonomous Region, China.
4. A slab-beam is a horizontal structural element capable of withstanding a load primarily by resisting bending. Often made of wood, it is used with purlins and rafters (diameter typically 20 cm or more).
5. CNY values in this table are based on a 2005 exchange rate of 8.3 CNY = $1. Currently the exchange rate is 6.3 CNY = $1.
6. The intensity of rainfall is a measure of the amount of rain that falls over time. The unit of rainfall intensity is usually millimetres per minute.
7. The big bell-mouthed stage is a growing stage of maize during which the length of the maize tassel spindle grows to about 0.8 cm.
8. A growing stage of wheat during which the wheat basal inter-node starts to elongate and grows above the ground by 1.5–2 cm.
9. A wheat growing stage during which 50 per cent of the plants in the field have the leaf sheath appearing on the tiller flag leaf and the ear enveloped by the flag leaf sheath inflates distinctly.
10. A wheat growing stage during which 50 per cent of the ears (awn not included) half appear from the leaf sheath.
11. A kind of pressurized irrigation system, similar to drip irrigation but that does not use drip emitters and operates at a higher water pressure and flow rate.
12. A check dam is a small dam, either temporary or permanent, built across a minor channel or drainage ditch. They reduce erosion and gullying in the channel and allow sediments and pollutants to settle. They also lower the speed of water flow during storm events.
13. Gully erosion occurs when water flows in narrow channels during or immediately after heavy rains or melting snow. The erosion is both downward, deepening the valley, and headward, extending the valley into the hillside.
14. A rafter is a type of beam that supports the roof of a building and is put perpendicularly on the purlin (diameter typically 4 cm).
15. A purlin is a horizontal structural element in a roof. Purlins support the loads from the roof deck or sheathing and are supported by the beam (diameter typically 8 cm).

References and further sources of information

References

China Ministry of Water Resources (2004) *Criteria for health evaluation of safe rural water supply*, Beijing.
http://wenku.baidu.com/view/6fbb425d3b3567ec102d8af6.html

China Bureau for Technical Supervision (1996) *China Regulation of Techniques for Comprehensive Control of Soil Erosion*, China Standard Press, Beijing

China Ministry of Water Resources (2001) *Technical Code of Practice for Rainwater Collection, Storage and Utilization*, China Water Resources and Hydropower Press, Beijing.

GBWR and GATS (Gansu Bureau of Water Resources and Gansu Administration of Technical Supervision) (1997) *Technical Standard of Rainwater Harvesting Project in Gansu*, Gansu Science and Technology Press, Lanzhou

GRIWAC (Gansu Research Institute for Water Conservancy), Dryland Agriculture Institute of Gansu Academy of Agriculture Sciences, Gansu Agriculture University, and Dingxi County Government (2002) 'Technology integration and innovation study on highly efficient concentration of natural rainfall in the semi-arid mountainous area', GRIWAC, unpublished report, Lanzhou.

Qian Zhengying, (ed.) (1991) *Water Conservancy in China*, China Water Resources and Power Press, Beijing.

Schumacher, E.F. (1973) *Small is Beautiful: A Study of Economics as if People Mattered*, Blond and Brggs, London.

Seckler, D. et al. (1998) *World water demand and supply, 1990 to 2025: Scenarios and issues*, International Water Management Institute (IWMI).

UNESCO, et al. (2006) *Water a Shared Responsibility, the United Nation World Water Development Report 2*, UNESCO Publishing and Berghahn Books

Zhu Q., Li, Y. and Ma, C. (2007) *Rainwater Harvesting*, Anhui Education Publishing House, Hefei.

Note: ***Since many of the references and unpublished research reports used in the compilation of this book are written in Chinese, only a few key publications have been cited above.***

Further sources of information

Gould, J. and Nissen-Petersen, E. (1999) *Rainwater Catchment Systems for Domestic Water Supply: Design, Construction and Implementation*, IT Publications, London (see Chapter 9 for a case study on China).

Zhu, Q. and Li, Y. (2000) 'Rainwater harvesting for survival and development', *Waterlines*, 18 (3), pp. 11–14.

Zhu, Q. and Li, Y. (2003) 'Drought-proofing villages in Gansu Province', *LEISA Magazine*, June, pp. 14–16.

Zhu, Q. and Li, Y. (2006) 'Rainwater harvesting: The key to sustainable rural development in Gansu, China', *Waterlines*, 24(4), pp. 4–6.

Zhu, Q. and Wu, F. (1995) 'A lifeblood transfusion: Gansu's new rainwater catchment systems', *Waterlines*, 14 (2), pp. 5–7.

Zhu, Q., Liu, C., Mou, H., Wang, H., Kung, H., Wanjun, Z., Zhijin, L. and Jiansheng, C. (2001) 'Rural water harvesting programmes, China: Weak water becomes a growth point' in Agrawal, A., Narain, S. and Khurana, I. (eds) *Making Water Everybody's Business: Practice and Policy of Water Harvesting*, pp. 170–82, Centre for Science and Environment, New Dehli, India, www.cseindia.org

Websites

CSE (Centre for Science and the Environment), www.rainwaterharvesting.org

Development Technology Unit, School of Engineering, University of Warwick, UK, www.eng.warwick.ac.uk/dtu/rwh

IRCSA (International Rainwater Catchment Systems Association) www.ircsa.org and IRCSA Fact Sheets www.ircsa.org/factsheets.htm

Practical Action www.practicalaction.org, www.practicalactionpublishing.org, and www.developmentbookshop.org